Anaerobic Digestion in Built Environments

Edited by Anna Sikora

Published in London, United Kingdom

IntechOpen

Supporting open minds since 2005

Anaerobic Digestion in Built Environments
http://dx.doi.org/10.5772/intechopen.92507
Edited by Anna Sikora

Contributors
Nadimkhandaker Khandaker, Mohammad Moshiur Rahman, De Salima Diba, Godwin Glivin, Premalatha Manickam, Mariappan Vairavan, Joseph Sekhar Santhappan, Martina Pilloni, Tareq Abu Hamed, Martha Osei-Marfo, Isaac Mbir Bryant, Alfredo de Jesús Martinez Roldán, María Dolores Josefina Rodriguez Rosales, Sergio Valle Cervantes, Thania Azucena Mendez-Perez, Jagannadh Satyavolu, Robert Lupitskyy

First published in London, United Kingdom, 2021 by IntechOpen
IntechOpen is the global imprint of INTECHOPEN LIMITED, registered in England and Wales, registration number: 11086078, 5 Princes Gate Court, London, SW7 2QJ, United Kingdom
Printed in Croatia

British Library Cataloguing-in-Publication Data
A catalogue record for this book is available from the British Library

Additional hard and PDF copies can be obtained from orders@intechopen.com

Anaerobic Digestion in Built Environments
Edited by Anna Sikora
p. cm.
Print ISBN 978-1-83969-223-9
Online ISBN 978-1-83969-224-6
eBook (PDF) ISBN 978-1-83969-225-3

For EU product safety concerns:
IN TECH d.o.o., Prolaz Marije Krucifikse Kozulić 3, 51000 Rijeka, Croatia, info@intechopen.com or visit our website at intechopen.com.

Meet the editor

Anna Sikora is a microbiologist, professor at the Institute of Biochemistry and Biophysics (IBB) PAS, Warsaw, Poland, and a leader of the Laboratory of White Biotechnology, Warsaw, Poland. She is a graduate of the Faculty of Biology, University of Warsaw. She received her Ph.D. and postdoctoral degrees from IBB PAS. She specializes in research on anaerobic digestion, fermentation processes, biogases (biohydrogen and biomethane) production; microbial communities and nutritional interactions between microorganisms; microbial iron reduction; and mutagenesis and DNA repair in bacteria. She has authored experimental papers, book chapters, and conference reports. She cooperates with the sugar industry in the field of applied research on the development of a method to obtain hydrogen and methane during anaerobic digestion of byproducts and wastes from sugar production.

Contents

Preface

Anaerobic digestion (AD) of biomass to methane and carbon dioxide is promoted by the activity and interactions of many different physiological groups of microorganisms that form specialized microbial communities. These microorganisms are responsible for four steps of AD: hydrolysis of polymeric organic matter to monomers, acidogenesis (acidic fermentations), acetogenesis (degradation of acidic fermentation products to dioxide, hydrogen, and acetic acid, which are substrates for methanogenesis), and finally methanogenesis, that is, the formation of methane. AD is a key process in the global carbon cycle and energy flow in ecosystems. It commonly occurs in natural anoxic ecosystems such as water sediments, wetlands, marshlands, and the digestive tracts of animals.

Biomethane and carbon dioxide are also generated in the environment as the result of human activity at landfill sites, anaerobic wastewater treatment plants, and biogas plants. Thus AD is an excellent method for utilization of wastes and production of green energy in local facilities located at small factories, workplaces, and in rural areas and housing complexes.

This book presents examples of AD solutions in specific regions and sites. Local installations are particularly important for producing dispersed energy and protecting the environment via the utilization of different types of waste such as textile processing wastewater, wastes of the pulp and paper industry, waste streams from grain processing, and domestic sewage. Through AD, energy and heat are supplied to nearby consumers, which has great economic significance. Furthermore, biogas-producing local installations contribute to increasing the share of renewable energy in overall energy production and to reducing greenhouse effects on a global scale.

Anna Sikora
Laboratory of White Biotechnology,
Institute of Biochemistry and Biophysics,
Polish Academy of Sciences,
Warsaw, Poland

Section 1

Processing of Industrial Wastewater

Chapter 1

Acetogenic Pretreatment as an Energy Efficient Method for Treatment of Textile Processing Wastewater

Nadim Reza Khandaker, Mohammad Moshiur Rahman and De Salima Diba

Abstract

This chapter will introduce the concept of a novel application of acetogenic pretreatment of textile processing wastewater. Acetogenic pretreatment is traditionally limited to high solids, easy to degrade wastewater to enhance degradation for methane generation. The application of the acetogenic process to a complex wastewater from textile processing facilities is novel and has the potential to remove color, chemical oxygen demand, biological oxygen demand in an energy efficient manner compared to the existing extended aeration processes applied in the industry. The application of the acetogenic process can be achieved to existing treatment facilities with minimum retrofit. The acetogenic operation will ensure the treatment process becoming greener with a small carbon footprint to achieve the goal of efficient wastewater treatment.

Keywords: Acetogenic, Pretreatment, Textile Processing Wastewater

1. Introduction

Anaerobic treatment of industrial wastewater was from its inception limited to wastewater that has a high concentration of biodegradable solids. The anaerobic biodegradation process is a multistep process where the first step in hydrolysis where extracellular enzymes secreted by microorganisms under anaerobic conditions solubilize the biodegradable solids, the subsequent steps being the conversion of the soluble organics in multiple steps to methane and carbon dioxide gas more commonly known as biogas [1]. The sequential transformation of solids to biogas is simply summarized in **Figure 1** below. It is important to note that the transformation process is sequential and complex, and more often than not, the rate limiting step determines the kinetics of the reaction and in most cases, this being the hydrolysis step where complex organics are broken down to soluble products such as organic acids alcohols, that are then converted to the common intermediator of acetic acid, which is then further transformed by methanogenic bacteria to methane and carbon dioxide [1–3].

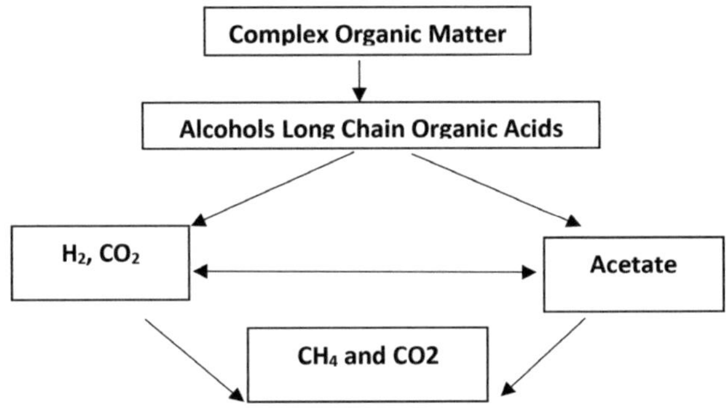

Figure 1.
Simplified schematic diagram of sequential transformation of organic compounds under the anaerobic condition to methane and carbon dioxide.

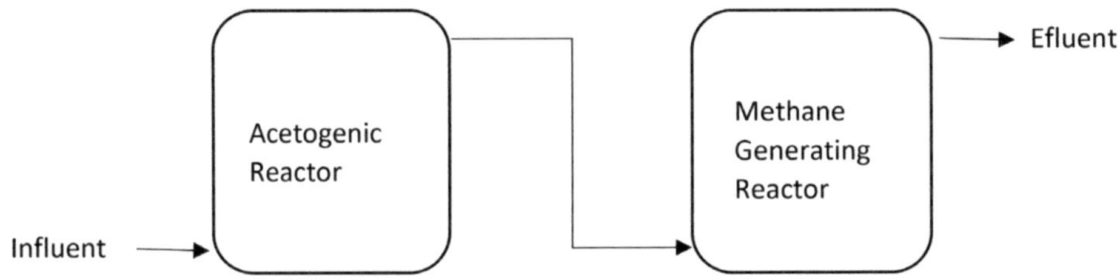

Figure 2.
Schematic of split transformation of acetogenic methanogenic reactor application.

In the era of sustainable development, the recent trends have been to optimize the process to enhance the production of biogas from biodegradable solids or high solids wastewater as a source of sustainable renewable energy, meaning biogas. Researchers have demonstrated that splitting the anaerobic process and its application as a two-step process and operating reactors as two-stage reactors (**Figure 2**). In the two-stage operation, the first stage reactor is followed by the methanogenic reactor to produce methane. The advantage of this split mode of operation is that more solids solubilized in the first step will increase the production of biogas in the second step. This split mode of application has been applied successfully to high solids wastewater, where the first reactor, referred to as acetogenic reactor, operates at a low hydraulic detention time in hours, generally between two to four days, followed by the methanogenic reactor with a hydraulic detention time of twenty days [4–6].

Acetogenic pretreatment has been limited in its application to high solids waste or wastewater in two-stage anaerobic reactors to enhance the hydrolysis of solids [7–9]. With a greener operation in mind, researchers have further progressed the acetogenic operation to optimized for hydrogen generation not just as a byproduct of gas of hydrolysis/acetogenesis but to produce hydrogen gas from organic waste solids. Hydrogen being a green fuel that can directly be used to generate electricity by using fuel cells [reference]. The thrust of the research has been to negate any methanogens in the acetogenic reactor, thereby increasing hydrogen yield. This chapter introduces the further progression of application of acetogenic operation of anaerobic reactors dedicated to the treatment of textile processing wastewater. At the laboratory level, progressive researchers have been applying the concept of the acetogenic process to pretreat textile processing wastewater in the hypothesis that anaerobic acetogenic operation of a reactor dedicated to textile wastewater will produce in the reduction of color, chemical oxygen demand, and total dissolved solids in an energy efficient manner [10].

2. Justification for application to textile wastewater

Textile wastewater is deleterious, containing complex organics, chroma, and also high in dissolved solids. If allowed to be realized to water bodies can be destructive to the aquatic environments. In recent decades the textile industries have been moving to developing economies to take advantage of the cheaper cost of production and deficiencies in regulatory requirements. Case in point Bangladesh, which is a developing industry and the second largest producer of readymade garments in the world. A forty-billion-dollar industry the largest employer of women and a progressive force that had bought the country from a least developed country to a middle income country in a few decades [11]. The flip side to all this is the negative impact on the environment of Bangladesh. Unabated discharge of untreated wastewater from the textile industries has severely affected the water bodies in the areas where the industries are located. The situation is so acute that in sections, the once ecologically sound rivers are highly polluted, and all aquatic life is dead. The picture below shows the unabated release of textile dye in a river in Bangladesh (**Figure 3**) [12, 13].

The reason more often than not for noncompliance by the industries is the cost of treatment [13, 14]. The convention wastewater treatment that is currently7 applied as the industry norm is chemically mediated settling to remove solids, the addition of decoloring agents to remove chroma, and extended biological activated sludge treatment (extended aeration with hydraulic detention times of greater than 13 hrs). The schematic flow diagram of the extended aeration chemically aided process currently used in Bangladesh and other countries is to treat textile processing wastewater shown in **Figure 4** below [11].

The extended aeration process is dependent on chemicals for the settling of solids and also chroma removal; the secondary biological treatment is energy intensive, requiring 7.0 kWh of energy per Kg of BOD_5 stabilized due to the aeration required by the aerobic microorganisms in the extended aeration process for operation of the blowers required for aeration. If we can negate the requirement of chemicals for chroma removal and solids removal and also reduce the BOD_5 loading to the secondary extended aeration system, this would call for a cheaper and energy efficient process and not to mention the reduction of greenhouse gas emission due to reduced energy requirements of the operation. Acetogenic pretreatment would provide an option of pretreatment that would remove color and solubilize solids and also reduce BOD_5 in the wastewater and thereby reduce the BOD_5 loading to the secondary aerobic treatment and reducing aeration requirement and thus savings in energy. The schematic of the proposed process retrofit using acetogenic pretreatment is shown in **Figure 5** below.

In the subsequent sections, the efficacy of the acetogenic pretreatment when applied to textile wastewater will be elucidated, along with the potential

Figure 3.
The picture shows the unabated release of textile dye in a river in Bangladesh.

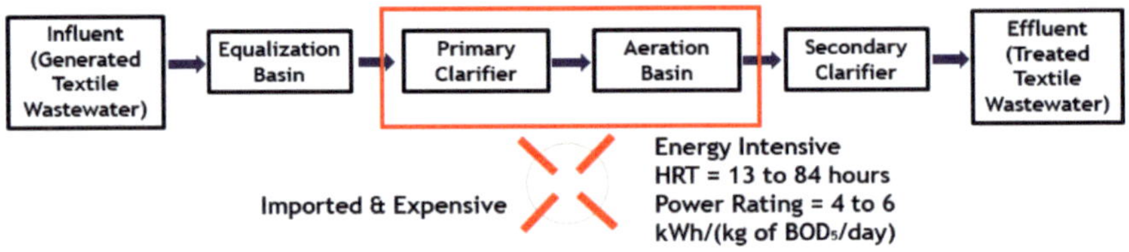

Figure 4.
The schematic flow diagram of the extended aeration chemically aided process currently used to treat textile processing wastewater.

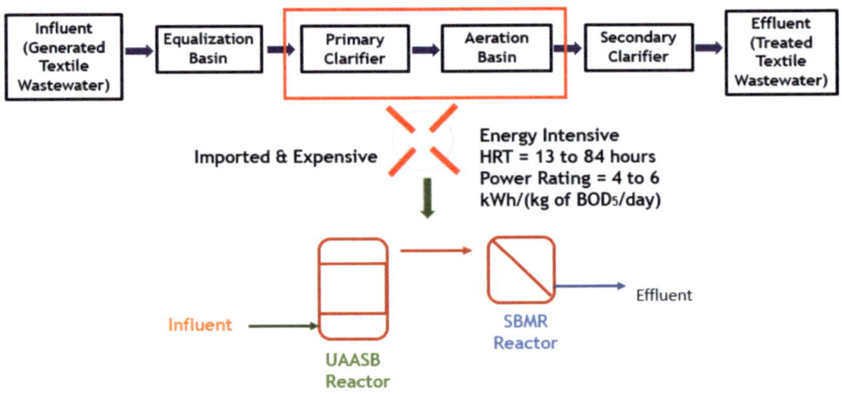

Figure 5.
The schematic of the proposed process retrofit using acetogenic pretreatment.

reduction in energy consumption in treatment will be highlighted. The discussion will be based on actual wastewaters from two textile processing industries.

3. Understanding acetogenesis as applied to textile processing wastewater pretreatment

The process of maintaining an acetogenic reactor is to curtail the growth of methanogenic microorganisms in a reactor and thereby stopping the conversion of fatty acid generated in the reactor to methane and carbon dioxide. Researchers reported controlling methanogenic microorganisms in an acetogenic reactor by shortening the hydraulic retention time greatly, usually at 2 to 4 days or even lower, in essence maintaining the reactor in a washout mode, thus limiting the growth of methanogens [15, 16]. Another method of controlling methanogenic microorganisms in an acetogenic reactor is oxygen shocked [17, 18]. Of the methods tried by prior researchers, the one that would be easier to apply in existing plants with minimum retrofitting. This method would be converting the existing basins such as equalization basins or parts of the extended aeration to the acetogenic reactor with retention time between 2-4 days and periodic shock aeration using existing in plant aeration capacities and equipment by nominal retrofitting [10]. To investigate this concept in the bench scale, acetogenic reactors were operated as proof of concept using actual textile processing wastewater. Two candidate wastewaters were used, one from a denim processing wastewater and another from com composite fabric processing wastewater. The findings of the bench scale study are summarized in the sub-headings below [10, 19].

3.1 Acetogenic reactor operation

The acetogenic reactors operated with the textile wastewaters were operated in a semi continuous batch mode with dally waste feeding; at a hydraulic retention time of 4.0 days, the reactor food to microorganism ratio (F/M) was constantly changing through the substrate loading was kept constant, thereby operating in a washout mode with a constant decrease of MLSS over the period of operation. This washout mode of operation ensures acetogenic conditions in the reactor operated under non forced aeration conditions. However, the periodic burst of shock aeration (dissolved oxygen raised to 2.0 mg/L once a day) to kill any growth of methanogenic microorganisms. The reactors used in this experimental program were flat bottomed class vessels in volume between 500 ml to 2000 ml. The test reactors were continually steered by means of a magnetic stirrer. The test reactors were plumed for sample withdrawal and feeding, along with plumbing and air diffuser systems for aeration. The aeration was provided using simple fish tank aerators through a fine air diffuser. The reactor vessel is insulated with temperature control. There is provision for temperature, dissolved oxygen, and pH monitoring in the reactor. The reactors were operated under the following conditions:

1. Maintained at mesophilic temperature (20-40°C),
2. A fixed hydraulic retention time,
3. A draw-and-fill waste feed schedule,
4. Waste feeding conducted once per day
5. No reactor pH adjustment, and
6. No augmentation of nutrients or buffer.

The raw textile wastewaters used in both the case studies reported in the following sections were obtained from textile processing facilities from the equalization basin. Time proportioned composite sampling procedure was used for the collection of the sample over a twenty-four-hour period of operation of the wastewater treatment plant [10, 19].

3.2 Acetogenic reactor seed source

The culture for the laboratory acetogenic cultures for both the case studies were from the sludge thickening tank that unaerated with a solids content of around 2%. The thickening tank contained waste activated sludge from the secondary clarifier of the extended aeration wastewater treating the complex wastewater in case study two discussed below.

3.3 Process operation parameters

The reactor per process operation parameters that were monitored were dissolved oxygen level during purging, reactor mixed liquor suspended solids, reactor pH, reactor temperature, reactor effluent color, total dissolved solids (for test case run 2), and reactor effluent chemical oxygen demand. Day two and day twenty reactor effluent sample for the second test case was sent for Furrier Transformation Inferred Spectroscopy.

3.4 Analysis procedures

The biological Oxygen Demand was measured using the serial dilution method HACH Method 8043, Chemical Oxygen Demand was measured by HACH method 8000 Digestion Method using preset vials 0-1500 mg/L rang, and the color was measured by HACH Method Platinum-Cobalt adapted from. Standard Method 8025 for the Examination of Water and Wastewater [20].

Total Dissolved Solids was measured using EC/TDS/NaCl probe and meter by HANNA Instruments HI 2300 system, and pH was measured HANNA HI 2211 pH/ORP probe and meter.

Total Suspended solids were determined by Standard Method 2540D, where a well-mixed volume of a sample was filtered through a pre-weighed glass fiber filter (pore size 0.45 micro meter). The filter was dried at 104°C and then weighed. The mass increase divided by the water volume filtered is equal to the Total Suspended solids (TSS) in mg/L [21].

The Fourier Transform Infrared (FTIR) spectrum of the reactor effluent was recorded using Bruker Vortex 70 FTIR. The spectra were taken in the range 400 to 4,000 cm^{-1}.

3.5 Case study acetogenic application to denim processing wastewater

The denim processing wastewater was characterized to have high total Chemical Oxygen Demand and high pH. The subject wastewater had a Chemical Oxygen Demand (COD) of 371 ± 37. mg/L, the color of 660 ± 66 ptco pH = 8.6 ± 0.6, and a five-day biological oxygen demand divided by the Chemical Oxygen Demand (BOD_5/COD) ratio of 0.62, indicating wastewater with a substantial organic fraction that should be biologically degradable. The wastewater was directly fed into the acetogenic reactor (Liquid volume 500 ml) without any adjustment, and the reactor operated as mentioned earlier in a waste feed more of semi-batch operation for a period of nine days. The results experimental program showed that after a period of acclimation, the acetogenic culture was able to completely remove the color and also produced substernal removal of chemical oxygen demand shown by respective parameters effluent concentrations decreasing with reactor operation (Refer to **Figures 5–8**). This clearly proved the efficacy of the process with ninety percent removal of color and greater than eighty percent removal of chemical oxygen demand for application for pretreatment of textile processing wastewater as an alternated to the chemical intensive decoloring and solids removal processes currently being employed. Reactor operating parameters also showed that beyond food to microorganism (F/M) operating ratio of 0.1, the system performance starts to decrease; this implies that for long term sustainable operation of acetogenic reactors, periodic reseeding with acclimated culture would be necessary [10]. Also, first order

Figure 6.
Picture of raw and treated wastewater showing clearly the efficacy of the process.

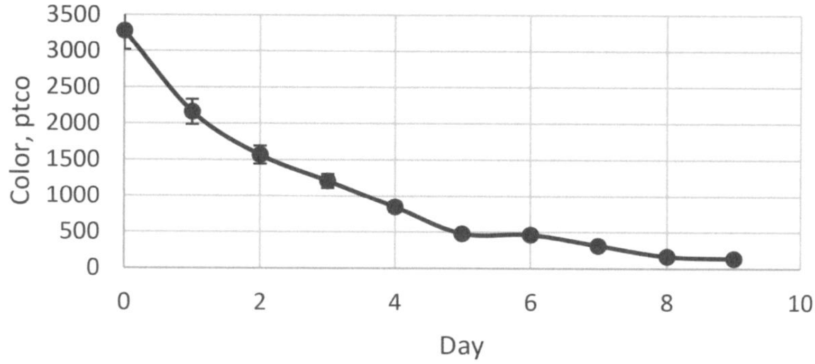

Figure 7.
Treated effluent color profile denim processing wastewater from the acetogenic pretreatment process for the denim processing wastewater.

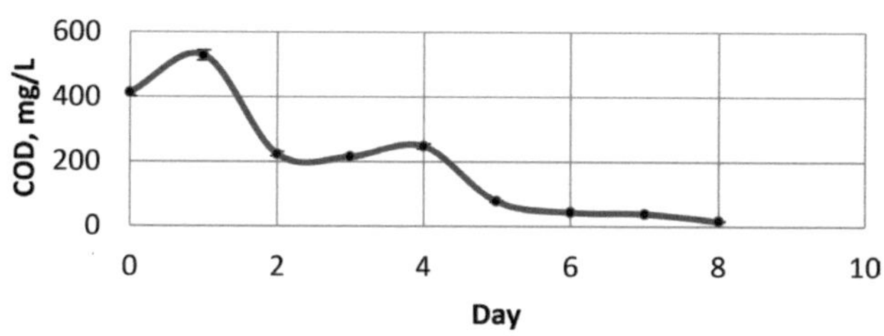

Figure 8.
Treated effluent Chemical oxygen demand profile denim processing wastewater from the acetogenic pretreatment process for the denim processing wastewater.

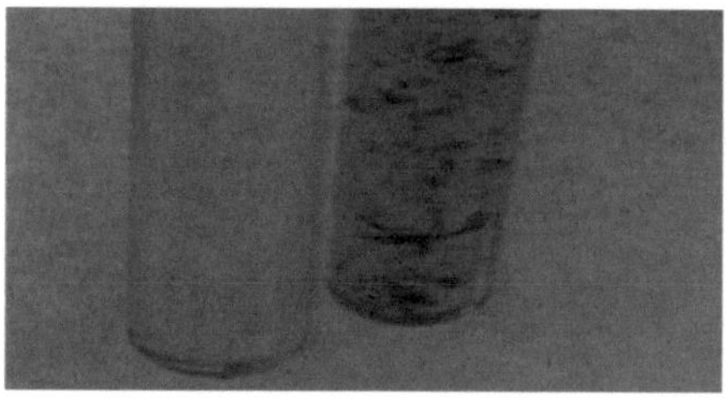

Figure 9.
Picture of raw and treated wastewater showing clearly the efficacy of the process for the complex wastewater.

rate kinetics defined both the color and chemical oxygen demand reduction and increased with days of operation and can be attributed to culture acclimation [10].

3.6 Case study acetogenic application to complex textile processing wastewater

The complex wastewater was from a composite factory where different fabrics are woven, dyed, textured, and finished stitched readymade garment products are produced. The facility that produces wastewater is varied and complex and was thought would be more of a challenging substrate to test the efficacy of the acetogenic process. Characteristics of the composite textile wastewater were color of 3540 ± 353 ptco, the chemical oxygen demand of 5186 ± 138 mg/L, BOD_5/COD ratio of 0.4, and pH of 9.6 ± 0.3 [19]. The proof of the efficacy of the acetogenic process in the treatment of textile processing wastewater is further illustrated in **Figure 9**, where the colloidal suspension is completely removed, indicating the extent of visual color removal. Furrier The acetogenic process was able to achieve for the complex textile processing wastewater with the color, and chemical oxygen removal was greater than 90 percent, along with a

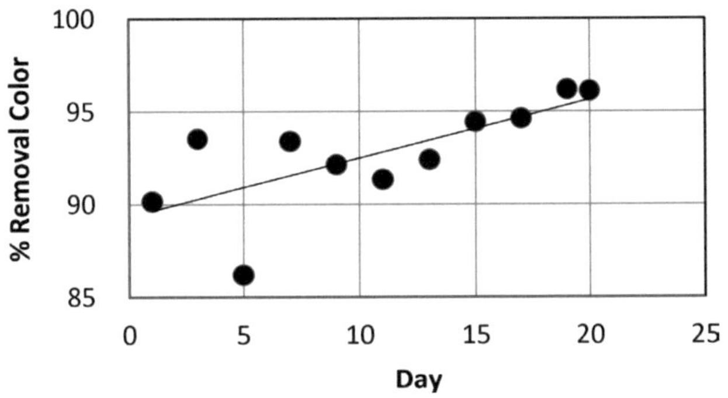

Figure 10.
Color removal efficiency for complex textile processing wastewater from the acetogenic pretreatment process.

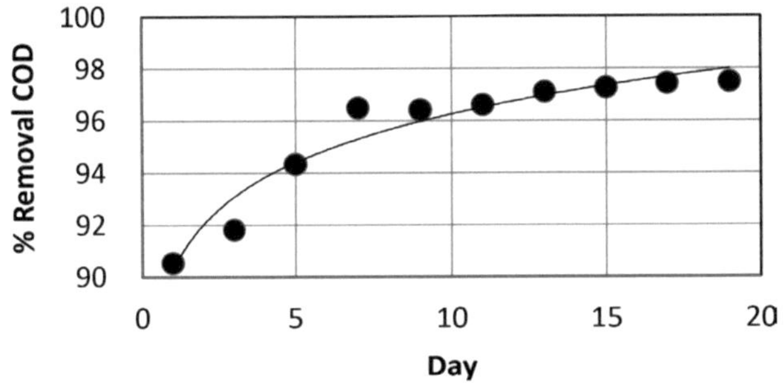

Figure 11.
Chemical oxygen demand removal efficiency for complex textile processing wastewater from the acetogenic pretreatment process.

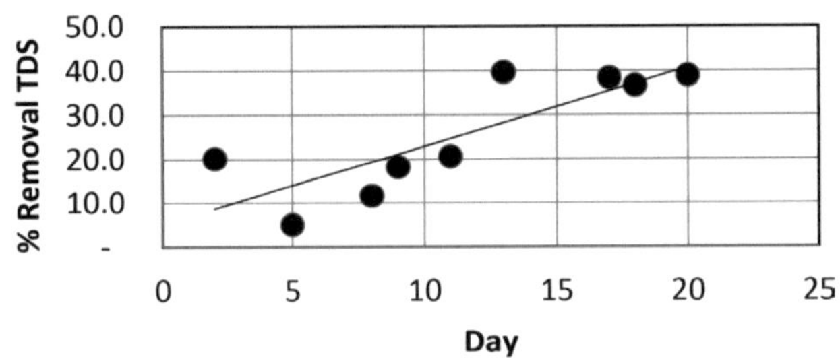

Figure 12.
Total dissolved solids removal efficiency for complex textile processing wastewater from the acetogenic pretreatment process.

reduction in total dissolved solids. The removal of total dissolved solids by the acetogenic process is an additional benefit as most textile processing wastewaters treated effluents have a hard time meeting the regulatory standers for total dissolved solids without employing expensive membrane systems [19]. **Figures 10–12** clearly illustrate the efficacy of the process with its high levels of color, Chemical Oxygen Demand, removal along with the removal of Total Dissolved Solids. Transformation Inferred spectroscopy further illustrates the efficacy of treatment where an effluent sample from day one (**Figure 13**) of the acetogenic reactor is compared to effluent from the acetogenic reactor on day 20 (**Figure 14**). The acetogenic reactor was operated for 20 days and the reactor operating liquid volume was 1000 m operated in a semi batch mode with daily waste feeding. The comparison clearly shows that with the reactor operating at a steady state prolonged operation, the complex organic peaks seen in the

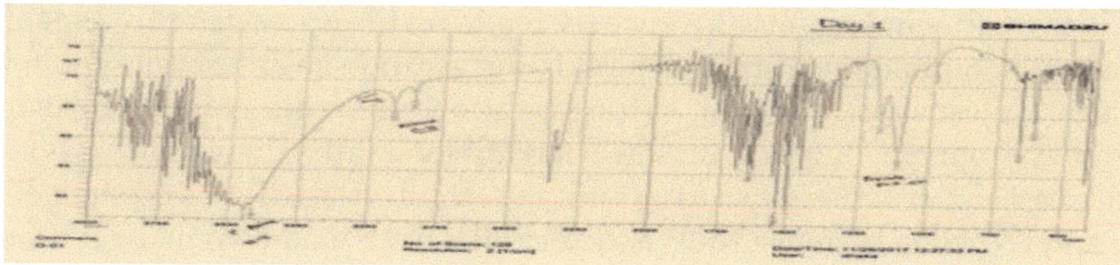

Figure 13.
Inferred spectroscopy effluent acetogenic reactor day 2.

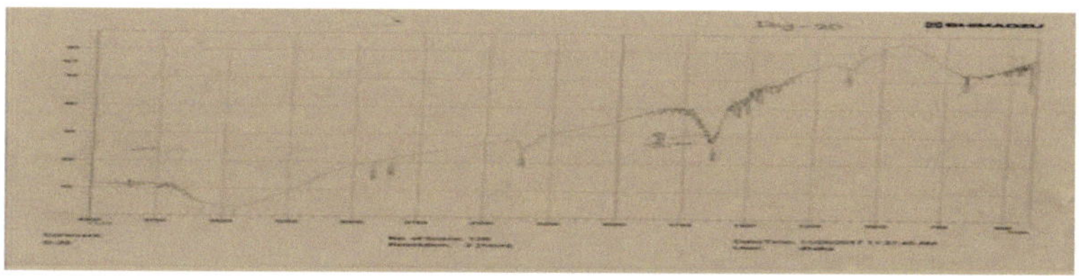

Figure 14.
Inferred spectroscopy effluent acetogenic reactor day 20.

effluent water were completely degraded by the acclimated acetogenic culture. They are again illustrating that the acclimated acetogenic culture can break down complex organics that are found in textile processing wastewater [19].

4. Energy savings potential and sustainable operation

Application of acetogenic pretreatment by reducing the biochemical oxygen demand/degradable chemical oxygen demand loading to the aerobic treatment system, which in return reduce the aeration volume, thus and reduce the electric energy requirement for running the aeration blowers. Aerator's energy consumption can range from 4.0 – 6.0 kWh/day-(kg of BOD_5/day) based on the type of blowers and aerators used [22]. Based on the database of the existing treatment plant wastewater characterization and laboratory study outlined for the case study one for the denim processing wastewater, the estimated energy requirement at a daily average flow of 722 m^3/day and the BOD_5 value of 228 mg/L the calculated BOD_5 loading to the existing aerobic basin at present is 164 kg of BOD_5/day. The plant uses fine bubble air diffusers with an energy rating of 4.0 kWh/day-(kg of BOD_5/ day energy requirement of 656 kWh/day. Based on the 85% BOD_5/COD removal efficacy of the acetogenic process, this would lead to loading of only 41 kg of BOD_5/ day and a blower energy requirement of 164 kWh/day, a net savings in energy of 495 kWh/day, a substantial saving of energy for any developing economy, case in point the energy requirement of an emerging economy like Bangladesh has a per capita annual energy requirement of 320 kWh [23].

5. Potential for industrial application

The acetogenic operation works when applied to pretreatment of textile processing wastewater for removal of color, reduction of COD, BOD5, and TDS. The process only requires periodic purging with air in contrast to the aerobic extended aeration process requiring constant aeration with substantial energy to operate the blowers. The proposed process can be applied to existing extended aeration wastewater treatment systems already existing in textile wastewater treatment facilities with nominal retrofitting. The existing aeration basin aerators could be modulated for just shock aeration, cutting aeration time from 24 hours a day to few minutes producing huge savings in electrical by limiting blower operation. The existing infrastructures also have built in secondary clarifiers and sludge storage and recycling systems; thus, added capital investment would be limited. It is anticipated that acetogenic pretreatment could be introduced with just process operational changes. Besides savings in energy, there would be a huge windfall in chemicals cost saving, for there would be no need for pH adjustment, activated carbon for color removal. All in all, acetogenic operation, with its reduced energy requirements and negating the needs of operating chemicals, makes it a greener viable option for textile wastewater treatment.

6. Conclusions

In an overall prospective the following conclusions can be drawn with regards to the application of acetogenic process to textile processing wastewater:

1. The acetogenic process can be applied to textile processing wastewater as a pre-treatment option to successfully remove color, chemical oxygen demand, total dissolved solids with a high degree of efficiency.

2. An added future of the process is that it requires periodic purge aeration rather than continuous aeration thus producing savings in energy for continuous functioning of aerators. Also the biological acetogenic process negates the requirement of chemicals for decoloring used in the conventional processes currently employed for textile wastewater treatment.

3. The seed culture requires a period of acclimation towards the treatment of textile wastewater and is operated in a washout mode with periodic seeding of recycled acclimated acetogenic culture.

4. The existing infrastructures can easily be retrofitted by modulating the aerators for shock aeration, and use the existing built in secondary clarifiers and sludge storage and recycling systems to reinject periodic acclimated culture to the acetogenic reactor for sustained operation.

Author details

Nadim Reza Khandaker*, Mohammad Moshiur Rahman and De Salima Diba
North South University, Dhaka, Bangladesh

*Address all correspondence to: nadim.khandaker@northsouth.edu

References

[1] McCarty, P. L. (1964). Anaerobic Waste Treatment Fundamentals. *Public Works*, 107-112.

[2] Young, J. C., & Tabak, H. H. (1993). Multilevel protocol for assessing the fate and effect of toxic organic chemicals in anaerobic treatment processes. *Water Environ Res, 65*, 34-45.

[3] Young, J. C., & Khandaker, N. R. (2000). Effect of Culture Acclimation on the Kinetics of Aldicarb Insecticide Degradation under Methanogenic Conditions. *Journal of Agricultural and Food Chemistry, 48*(4). doi: 10.1021/jf990427t

[4] Liu, T., & Ghosh, S. (1997). Phase separation during anaerobic fermentation of solid substrates in an innovative plug-flow reactor. *Water Science and Technology, 36*(6-7), 303-310.

[5] Mata-Alvarez, Macé, S., & Llabrés, P. (2000). Anaerobic digestion of organic solid wastes. An overview of research achievements and perspectives. *Bioresource Technology, 74*(1), 3-16.

[6] Pavan, P., Battistoni, P., Cecchi, F., & Mata-Alvarez, J. (2000). Two-phase anaerobic digestion of source sorted OFMSW (organic fraction of municipal solid waste): performance and kinetic study. *Water Sci Technol, 41*(3), 111-118.

[7] Bo, Z., & Pin-jing, H. (2013). Performance assessment of two-stage anaerobic digestion of kitchen wastes. *Performance assessment of two-stage anaerobic digestion of kitchen wastes*, 1277-1285.

[8] F. Baldiai, Pecorini, I., & Iannelli, R. (2019). Comparison of single-stage and two-stage anaerobic co-digestion of food waste and activated sludge for hydrogen and methane production. *Renewable Energy*, 1755-1765.

[9] Zuoa, Z., Wu, S., Qi, X., & Dong, R. (2015). Performance enhancement of leaf vegetable waste in two-stage anaerobic systems under high organic loading rate: Role of recirculation and hydraulic retention time. *Applied Energy*, 279-286.

[10] Sarker, M. S., Rahman, M. M., & Khandaker, N. R. (2018, May). Defining the kinetics of the novel application of anaerobic acetogenics for treating textile dyeing wastewater. *Earth Syst. Environ., 4*.

[11] Khandaker, N., & Talha, M. A. (2016). The new nexus and the need for understanding textile wastewater treatment. *BUFT Journal*, 19-26.

[12] Hossain, L., Sarker, S. K., & Khan, M. S. (2018). Evaluation of present and future wastewater impacts of textile dyeing industries in Bangladesh. *Environmental Development*, 23-33.

[13] Sakamoto, M., Ahmed, T., Begum, S., & Huq, H. (2019). Water Pollution and the Textile Industry in Bangladesh: Flawed Corporate Practices or Restrictive Opportunities? *Sustainability*.

[14] Khandaker, N. R., Afreen, I., Samina, D. D., Huq, F. B., & Akter, T. (2020a). Treatment of textile wastewater using calcium hypochlorite oxidation followed by waste iron rust aided rapid filtration for color and COD removal for application in resources challenged Bangladesh,. *Groundwater for Sustainable Development*.

[15] Kotsopoulos, T. A., Fotidis, I. A., Tsolakis, N., & Martzopoulos, G. G. (2009). Biohydrogen production from pig slurry in a CSTR reactor system with mixed cultures under hyper-thermophilic temperature. *Biomass and Bioenergy*, 1168-1174.

[16] Ramos, L. R., de Menezes, C. A., Soares, L. A., Sakamoto, I. K., Varesche, M. B., & Silva, E. L. (2020). Controlling methane and hydrogen production from cheese whey in an EGSB reactor by changing the HRT. *Bioprocess and Biosystems Engineering*, 673-684.

[17] Chae, K.-J., Choi, M.-J., Kim, K.-Y., Ajayi, F. F., Park, W., Kim, C.-W., & Kim, I. (2010). Methanogenesis control by employing various environmental stress conditions in two-chambered microbial fuel cells. *Bioresource Technology*, 5350-5357.

[18] Rafieenia, R., Pivato, A., & Lavagnolo, M. C. (2018). Effect of inoculum pre-treatment on mesophilic hydrogen and methane production from food waste using two-stage anaerobic digestion. *International Journal of Hydrogen Energy*, 12013-12022.

[19] Khandaker, N. R., Rio, F. F., Sarkar, L., & Sharmin, A. (2020b). Acetogenic Aerobic Sequential Batch Reactors in Series Operation for Textile Wastewater Treatment. *Journal of Environmental Treatment Techniques*, 766-769.

[20] HACH. (2013). *Calorimetr Procedure Manual.* Denver: HACH Chemical Company.

[21] American Public Health Association/American Water Works Association/Water Environment Federation. (1998). *Standerd Methods for Examination of Water and Wastewater.* New York: APH-AWWA-WEF.

[22] Eckenfelder, W. W., Melina, J. F., & Patterson, J. W. (1992). *Acivated Sludge Process Design and Control.* Lancaster: Technomic Publishing CO.

[23] Khandaker, N., Sarker, M., & Rahman, S. D. (2017). Acetogenic Pretreatment of Textile Wastewater for Energy Conservation. *International Journal of Science and Research*, 423-426.

Chapter 2

Evaluation of Physical and Chemical Pretreatment Methods to Improve Efficiency of Anaerobic Digestion of Waste Streams from Grain Processing

Jagannadh Satyavolu and Robert Lupitskyy

Abstract

Globally, Anaerobic Digestion (AD) industry is booming and biogas, the most sustainable biofuel, produced via AD is in an exponential market growth curve. According to a November 2020 report from US Energy Information Administration (EIA), "25 large dairies and livestock operations in the United States produced a total of about 224 million kWh (or 0.2 billion kWh) of electricity from biogas". However, the growth of AD and the cost-effective use of the generated biogas are hindered by the inconsistencies (composition, suspended solids, flow rate, etc.) of the incoming waste stream and the associated biogas quality (due to the presence of hydrogen sulfide gas). A pretreatment step prior to an AD unit can promote consistency in the incoming stream, minimize the suspended solids; and thereby insures the efficiency of AD. In this study, we evaluated the method of pretreatment of waste streams from three grain processing industries, where 1) we adjusted the pH of a stream corresponding to its isoelectric point (zero zeta-potential), 2) removed solids (and their corresponding COD) that precipitated, and 3) produced a consistent composition stream to feed the AD process. For grain processing industry, the precipitated solids can be returned to their process – thus integrating the pretreatment with the rest of the process. The pH pre-treatment should not add any additional cost to the plant since the pH of the waste streams from grain processing plant needs to be raised per plant permits prior to disposal. Our lab and pilot AD studies showed a positive effect of such pretreatment on these waste streams in terms of increased biogas production (11–60%) and COD removal (12–60%), and in some instances reduction in H_2S content in biogas (8%). This study clearly demonstrated that such a pretreatment method is economical and is effective to improve AD performance on waste waters from grain processing industries.

Keywords: anaerobic digestion, biogas, wastewater treatment, pH adjustment, grain processing

1. Introduction

Handling and treatment of industrial waste water has become one of the biggest problems of the last century due to constantly increasing industrial activity [1]. The amount of the industrial waste water is rapidly exceeding the biological treatment capabilities of the natural ecosystems. Hence, the treatment of industrial effluents became an important topic.

Anaerobic digestion (AD) is potentially an efficient and economically beneficial method of neutralization of industrial waste [2, 3]. Although anaerobic treatment was known for a long time, the process has not been successfully implemented owing to disadvantages, such as low sludge activity, low reactor capacity, unsuitability of the process and inhibitory effects [4]. The introduction of modern reactor designs where hydraulic retention time is uncoupled from the solids retention time led to a world-wide acceptance of the anaerobic technology as a cost-effective alternative to conventional waste water treatment methods. A number of reactor configurations have been developed leading to high biomass concentrations, such as upflow anaerobic sludge blanket (UASB) reactor, anaerobic contact filter, down flow stationary fixed film and anaerobic fluidized bed reactor (AFBR) systems [5]. In AFBR reactors, the sludge granules are fluidized by high up-flow fluid velocities generated by a combination of the influent and recirculated effluents. The fluidized bed process claims various potential advantages over other high rate anaerobic reactors [6]. These are: high sludge activity, high treatment efficiency, no clogging of reactors, no problems of sludge retention, least chance for organic shock loads and gas hold up as well as small area requirements. Currently, this anaerobic technology removes 70–90% of organic pollutants (expressed as chemical oxygen demand, COD).

In order to ensure high efficiency and high throughput of wastewater treatment using AFBR reactors, certain parameters, such as suspended solids, fat-oil-and-grease, complex organics (fiber, proteins), toxic compounds, should be minimized [7]. Pretreatment of industrial wastewater using physical and chemical methods can significantly improve efficiency of wastewater treatment using anaerobic technology [8]. One immediate impact of these pretreatments on the operation of an anaerobic digester is that its hydraulic retention time (HRT) can be lowered. HRT directly impacts the tank volume of the AD (capital cost) as well as the throughput from the digester. Hence the pretreatment methods can not only lower the capital cost of the anaerobic digestion, but also impact its operating cost.

Various physico-chemical pretreatment methods have been used to improve the anaerobic digestibility of the industrial waste streams. Filtration is used to decrease COD content, remove suspended solids, and toxic compounds [9, 10]. Enzymatic pretreatment is often used to improve digestibility of waste streams with high lipids content, such as dairy wastewater [11, 12]. Oxidative treatment with ozone is used to remove toxic organic compounds from the waste stream and improve anaerobic digestion [13]. Electrochemical treatment is often used for the destruction of recalcitrant organics and increase BOD_5/COD ratio [14–16]. pH adjustment has also been successfully implemented for various purposes as a pretreatment method. pH adjustment using $Ca(OH)_2$ was used to force ammonia stripping [17]. pH adjustment was also done to improve sludge dewatering after AD [18]. Alqaralleh [19] demonstrated the use of alkaline pretreatment to enhance the solubility of organics in the waste prior to AD. pH adjustment as a pretreatment method was also employed to precipitate proteins from wastewater [20, 21]. In another work, Cui and Jahng [22] removed proteins from disintegrated waste sludge prior to anaerobic digestion using pH adjustment to the corresponding isoelectric point (IEP) of the proteins.

Control of pH is a key operating parameter during anaerobic digestion process. However, industrial effluents very often have a pH that is not suitable for discharge or further processing. Hence pH adjustment of the waste stream to the discharge permit levels is done as an operating procedure prior to discharging the stream to further treatment. If, on the other hand, pH adjustment to bring the pH close to IEP can also serve as a pretreatment method, then we can reduce solids and other organics loading in the stream. This reduction will benefit a waste treatment process such as AD. Further, as discussed above, this pretreatment will not add any extra cost to the plant.

Solubility of many compounds depends on the IEP of the solution. Depending on the type of material being precipitated by adjusting to IEP, several advantages can be gained, such as decrease in COD, toxic compounds, complex organics, sulfates etc. This can lead to improved digestibility of the wastewater, as well as increased quality of the biogas [22, 23]. Delgenès et al. studied changes in anaerobic digestibility of industrial microbial biomass after thermochemical pretreatment. It was determined that the observed poor biodegradability and biotoxicity of the solubilized microbial biomass is due to high molecular compounds (>100 Da). Removal of these compounds using absorbent resins and precipitation by pH adjustment improved the biogas production. An increase in biogas production and biogas quality was observed as a result of the deproteination using pH adjustment to IEP [22]. In our study, we used pH adjustment to bring zeta-potential of waste streams from grain processing industries, such as distillery, soy protein processing, and oat fibers processing to near IEP as a pretreatment method. The objective is to reduce organic and solutes loading in the stream and thereby improve COD reduction, biogas yield and quality during anaerobic digestion of the waste streams.

2. Materials and methods

2.1 Materials

Calcium chloride, magnesium chloride, ammonium chloride, potassium phosphate monobasic, sodium sulfate were used as minerals and nutrients for anaerobic digestion tests and were purchased from Sigma-Aldrich. Sodium bicarbonate (Sigma-Aldrich) was used to adjust alkalinity. A proprietary inorganic salt mix (Respirometer Systems & Applications LLC, Fayetteville, AZ, USA) was used as a source of trace elements. Sodium hydroxide and hydrochloric acid (Sigma-Aldrich) were used for pH adjustment. Ethanol was purchased from Sigma-Aldrich and was used as a model source of COD. Granular anaerobic sludge was kindly provided by Anheuser-Busch (St. Louis, MO). The concentration of the bacteria in the sludge was measured as Volatile Suspended Solids (VSS) content and was determined to be 52.0 g/L.

2.2 Anaerobic digestion tests

Experimental set-up for laboratory-scale batch anaerobic digestion tests was acquired from Respirometer Systems & Applications LLC, Fayetteville, AZ, USA, and is shown in **Figure 1A**. It consists of a water bath placed on a 8-position magnetic stir plate, external pump and temperature controller, and a pulse flow respirometer PF-800. 500 ml glass bottles were used as reactors. Up to 8 bottles can be accommodated in the water bath. Trace elements, minerals, nutrients, and $NaHCO_3$ were added to each bottle as described elsewhere [24]. Substrates were added to the bottles in the predetermined amount so that the COD load was the same in each bottle. Bottles were inoculated with granular anaerobic sludge in the quantity so that the ratio between the substrate (expressed as mg/L COD) and the anaerobic bacteria (expressed as mg/L VSS) was 1:2. Bottles with ethanol substrate were used as a control. Ethanol is quickly and easily digested by methanogenic archaea and is

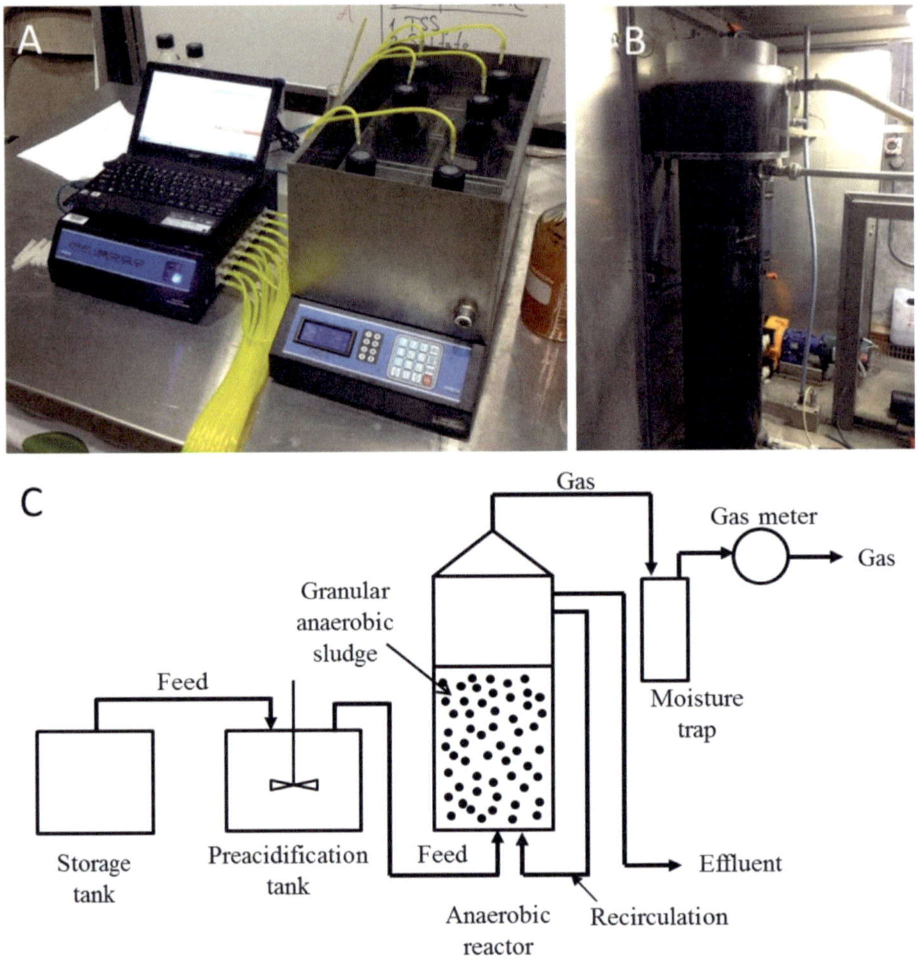

Figure 1.
Experimental set-up used for anaerobic digestion tests: (A) laboratory-scale batch unit; (B) continuous pilot-scale unit (main reactor only), and (C) block-scheme of continuous pilot-scale unit.

therefore used as a benchmark for substrate digestibility [24]. The pH after adding the biomass, substrates, and nutrients was 7. The bottles were degassed with nitrogen for 1 min to ensure anaerobic conditions. The anaerobic digestion tests were conducted under mesophilic conditions (35 °C). The volume of the biogas produced was measured and recorded by the pulse flow respirometer. The test was conducted for two feeding cycles. Each feeding cycle constitutes a reaction time frame during which all nutrients are consumed and gas production stops. After the first feeding cycle ends, the nutrients are replenished and the second feeding cycle starts. For each following feeding cycle, the biomass in the bottle was not removed or added. All the lab tests were performed in duplicate. These lab tests are done prior to pilot tests in order to evaluate the activity of the biomass for each of the streams, digestibility, and biogas quality. The lab tests helped us to better plan and design pilot tests.

Pilot-scale anaerobic digestion tests were performed on 60 L 2-stage Anaerobic Fluidized Bed Pilot Reactor (Voith Meri Environmental Solutions Inc., Appleton WI) shown in **Figure 1(B** and **C)**. In this reactor design, acidogenic and methanogenic stages are spatially separated: acidogenesis occurs mainly in the preacidification tank and the methane formation happens in the main reactor. It is designed to optimize the methane formation. First, the waste water is pumped from a 10 gallon storage tank into the preacidification tank, where it is kept until the acidification degree (ratio between volatile fatty acids content and COD content) reaches approximately 30% (**Figure 1C**). Then, the acidified wastewater is fed into the main reactor from the bottom, where granular anaerobic sludge resides. The stabilized wastewater is recirculated back at the 200 l/h rate. The recirculation is required to fluidize the granular sludge bed. The excess of the stabilized wastewater (effluent) is removed via the overflow channel and discarded. The gas is collected from the top of the reactor and, after passing through the moisture trap and gas meter, is discharged into the exhaust pipe. The reactor was inoculated with 40 L of anaerobic granular sludge. Each test was conducted for a 2-week period. Samples were taken on a daily basis and analyzed. The reactor was maintained at COD load of 3.0 ± 0.2 g-COD/L/day (feed rate 0.75 l/h; HRT 80 hours). The temperature in the preacidification tank and the main reactor was maintained at 36 ± 3 °C. The pH in the preacidification tank was automatically maintained at 5.5 by dosing NaOH. The pH in the main reactor was self-maintained at 6.8.

AD at lab and pilot scale was evaluated on at least two types of streams for each waste water type - a control (no pH adjustment) sample and a pretreated sample. Repeats and additional tests are conducted as needed. The data presented is a compilation of the multiple runs for each stream.

2.3 Analytical methods

Chemical analysis of the waste water was performed spectrophotometrically using commercial test kits and DR 3900 Spectrophotometer (Hach Company, Germany). Gas analysis was performed on SRI 8610C Gas Chromatograph (SRI Instruments Inc., Las Vegas NV) using HayeSep D column (Restek Corporation) and thermal conductivity detector (TCD) for methane and carbon dioxide detection; MXT-1 column (Restek Corporation) and flame photometric detector (FPD) was used for hydrogen sulfide detection. Z-potential measurements were performed on 90 Plus Particle Size Analyzer (Brookhaven Instruments Corporation, Holtsville NY).

3. Results and discussion

3.1 Wastewater characterization

Three types of wastewater streams from local grain processing industries have been used in our experiments: distillery, soy protein processing, and oat fiber processing. These streams have been analyzed for their chemical composition, physico-chemical properties, and solids content (**Table 1**). Samples from each operation were received 3–4 times a week for a three-week period in order to assess variability in the wastewater content. Therefore, some of the data in the table are presented as a range, representing the amplitude of variation of a particular parameter.

The solids in the distillery waste stream were separated by centrifuging at 1000 rpm for 15 min. The resulting liquor had a suitable mineral composition: sufficient nitrogen and phosphorus content and low sulfates. Soy protein processing wastewater had suitable COD content, low suspended solids, sufficient nutrients, but had very high sulfates content, which was in the range of toxicity for methanogenic archaea [25, 26]. Oat fiber processing wastewater had a high COD content, suitable mineral composition, but had a very high initial pH.

		Distillery	Soy proteins	Oat fibers
Electrochemical analysis	pH	3.9–4.6	4.0–4.2	11.3
	Isoelectric point	6.1	5.2	—
	Conductance, mS	6.7–7.1	10.3	22.0
Solids	Total solids, g/L	58.2–62.1	22.4–26.1	72.5
	Total suspended solids, g/L	32.7–34.9	3.2–5.3	11.5
	Total dissolved solids, g/L	25.1–27.1	19.7–21.5	61.0
Oxygen demand	Total COD, mg/L	53,600-57,200	16,500-18,000	85,000
	Soluble COD, mg/L	28,000-33,000	14,500-17,600	72,000
Chemical analysis	Sulfates, mg/L	129–256	4,400-5,500	300
	Phosphates, mg/L	40–226	74–106	300
	Ammonia, mg/L	20–50	44–69	40
	TKN, mg/L	28.9–30.1	37.8–42.3	n/d

Table 1.
Summary of the wastewater characterization.

3.2 Wastewater pretreatment

All three waste streams have initial pH that is not suitable for anaerobic digestion, which should be in the 6.5–7.5 range. Distillery and soy protein processing waste streams come at pH 3.9–4.6, which is too low, whereas oat fiber processing waste stream has pH of 11.3, which is too high. Adjusting pH prior to anaerobic treatment not only ensures the proper conditions for methanogenic archaea, but also makes the stream more consistent, eliminating any possible upsets in the AD reactor. Yet another advantage of pH adjustment is the possibility to precipitate colloidal solids by bringing the system close to its isoelectric point. We studied the pH-induced precipitation in these streams by changing pH in increments from 0.5 to 1.0 and measuring the zeta-potential as a function of pH to determine the IEP of the stream (**Figure 2**). Sodium hydroxide and hydrochloric acid were used for pH adjustment throughout the study. For distillery and soy protein processing streams, the pH-induced precipitation was studied in the range from original pH (~4) until 9. For both streams a precipitation was visually observed upon reaching pH of ~6.0 and ~ 5.4 for distillery and soy protein processing streams respectively. The extent of precipitation as a function of pH was studied by measuring COD at different pH points (**Figure 3**) after the sample has been centrifuged at 4000 rpm for 15 min. The highest decrease in COD content was observed at pH ~7 for distillery sample (6.5% COD decrease) and at pH ~6 for soy protein processing sample (10.3% COD decrease). Both points of highest COD decrease are either close or within the range of optimal pH for anaerobic digestion. It is noteworthy that these pH points are in the vicinity of the corresponding isoelectric points measured for these waste streams (**Figure 2A** and **B**). This suggests that the precipitated material is most likely a fraction of water soluble proteins. Oat fiber processing waste stream also showed pH-induced precipitation. In this case pH was reduced gradually from original pH of 11.3 to 2. After pH was decreased below 5, a significant precipitation was visually observed. The graph in **Figure 1C** shows pH-dependent COD decrease

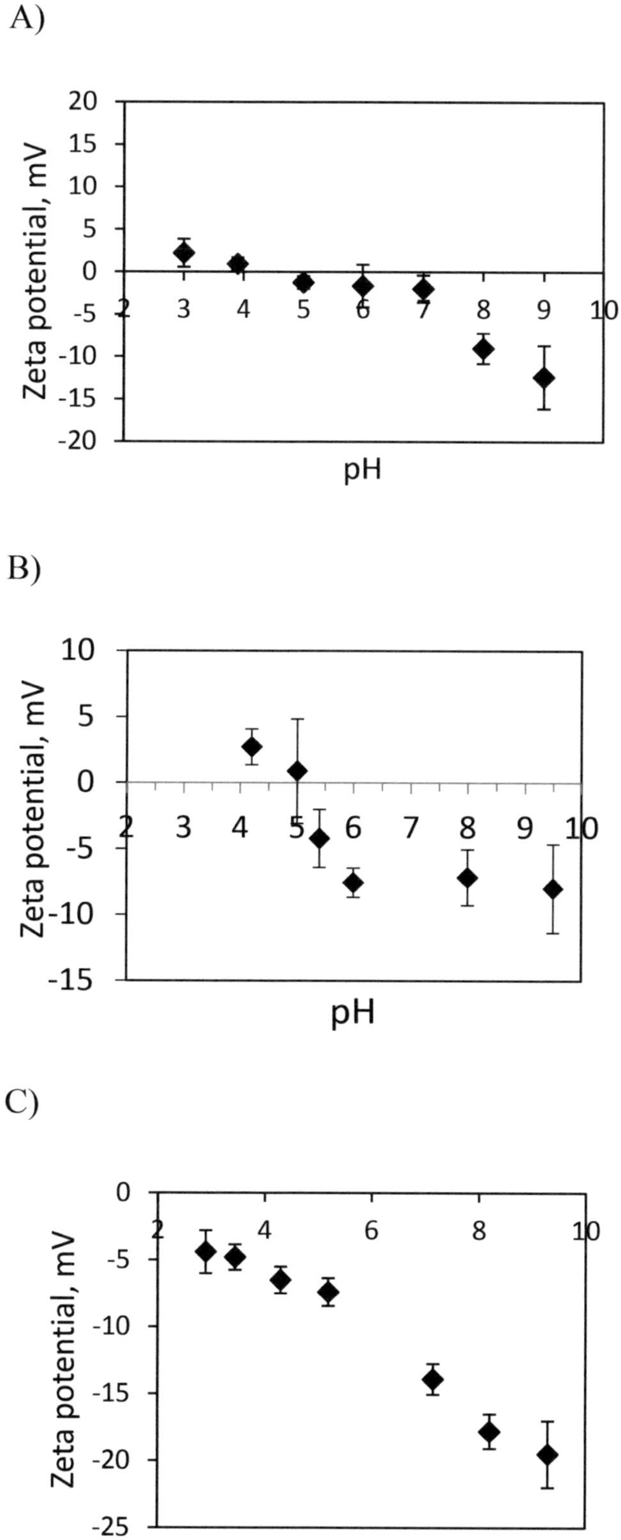

Figure 2.
Z-potential of the waste stream from (A) distillery, (B) soy protein processing, and (C) oat fiber processing as a function of pH.

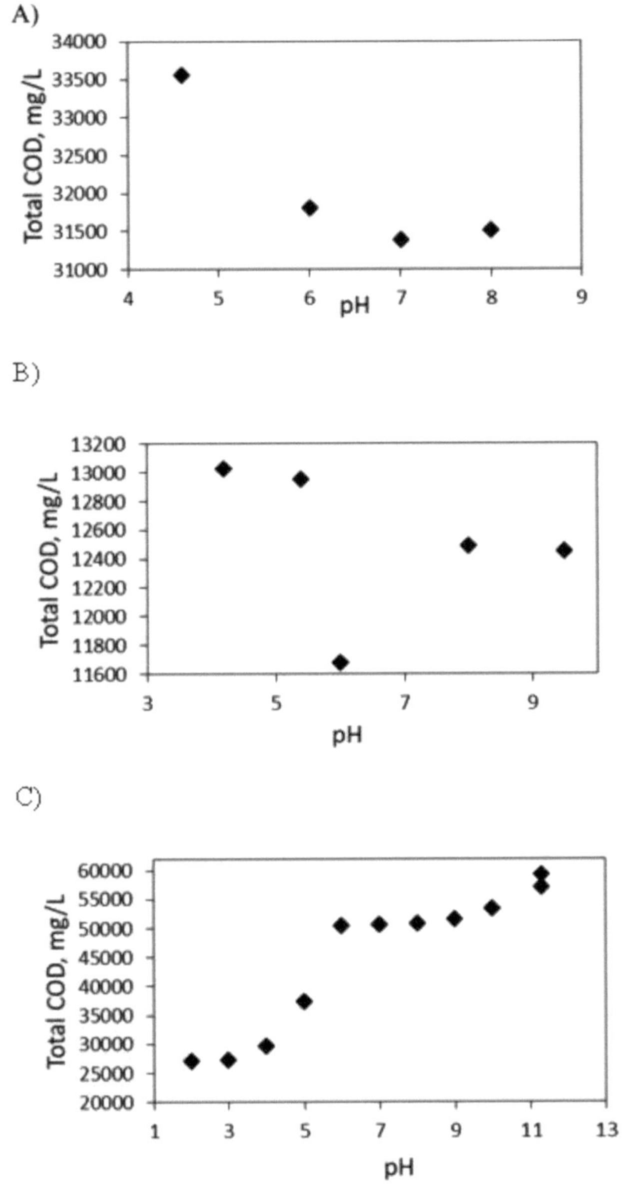

Figure 3.
Total COD as a function of pH of the wastewater from (A) distillery, (B) soy protein processing, and (C) oat fiber processing.

for this waste stream. A slight decrease in COD is observed as pH decreases from 11.3 to 6, followed by a rapid decrease in the pH range from 5 to 3. Overall, adjusting pH from 11.3 to 3 resulted in the removal of nearly 50% COD. Constant increase in precipitation throughout the entire pH range studied, combined with no isoelectric point in this range (**Figure 2C**) suggests that the precipitated material is most likely an alkali-soluble polycarbohydrates.

We also studied changes in the mineral composition of the waste streams upon pH adjustment (**Table 2**). Removal of dissolved solids upon pH adjustment in the soy protein processing wastewater resulted in the decrease of sulfates content by 16% and phosphates by 11%. Reduction of sulfates concentration is beneficial because

	Distillery		Soy protein processing		Oat fiber processing	
	Non-pretreated	Pretreated	Non-pretreated	Pretreated	Non-pretreated	Pretreated
pH	4.6	7.0	4.2	6.0	11.3	3.0
Sulfates, mg/L	134.2	131.4	4430.3	3710.7	306.2	304.4
Phosphates, mg/L	63.7	61.9	75.6	67.2	300.0	290.4
Ammonia, mg/L	26.4	25.9	48.4	46.8	40.1	38.8

Table 2.
Changes in the chemical composition of the wastewater upon pH adjustment.

high concentration of sulfate ions cause sulfide toxicity during anaerobic digestion process [25] Ammonia content did not decrease significantly. The above minerals in the other two waste streams did not change noticeably upon pretreatment.

3.3 Batch anaerobic digestion tests

We performed a laboratory-scale batch anaerobic digestion study in order to evaluate the effect of pretreatment on the anaerobic digestion of the wastewater in terms of biogas production, its quality, and possible inhibitory effects on the biomass activity. Pretreatment of the waste streams was performed by adjusting pH to the value that resulted in maximum decrease of COD content (**Figure 3**). Thus, the pH of the distillery and soy protein processing streams was adjusted to 7 and 6 respectively. The pH of the oat fiber processing waste stream was first adjusted to 3 to induce precipitation and, after removal of the precipitate, the pH was increased to 6 to bring it within the range suitable for methanogenic archaea. In all AD tests, separation of the precipitated solids was performed by carefully decanting the liquid after the precipitate was allowed to settle.

3.3.1 Distillery wastewater

Results of batch digestion test for the distillery wastewater before and after pretreatment are summarized in **Figure 4**. The experiment was conducted for two feeding cycles. Cumulative biogas production over each feeding cycle is presented in **Figure 4A** and corresponding specific methane production is shown in **Figure 4B**. For both feeding cycles, a clear increase in gas production is observed from the pre-treated sample. The total biogas production from the pretreated sample after 40 hours of digestion was 18% and 11.5% higher for 1st and 2nd feeding cycle respectively, compared to the non-pretreated sample (**Table 3**). As a result of pretreatment, COD reduction during the second feeding cycle increased from 80.2% to 89.4% (compared to control). Analysis of biogas samples (**Table 3**) indicated a slight decrease (8%) in H_2S concentration after the pretreatment, which may be due to the removal of the fraction of soluble proteins upon pH adjustmet. Protein-rich streams are known to have increased levels of H_2S in biogas [27]. In addition, corn gluten is particularly rich in sulfur-containing aminoacids, compared to other seeds [28]. The biogas composition, presented in **Table 3** and subsequent tables, does not add up to 100%, because biogas contains other minor components (typically hydrogen, nitrogen, oxygen, and moisture). Since, the emphasis of the study was on COD conversion, biogas production, and methane content as a function of pretreatment, elucidation of the complete biogas composition was beyond the scope of this manuscript.

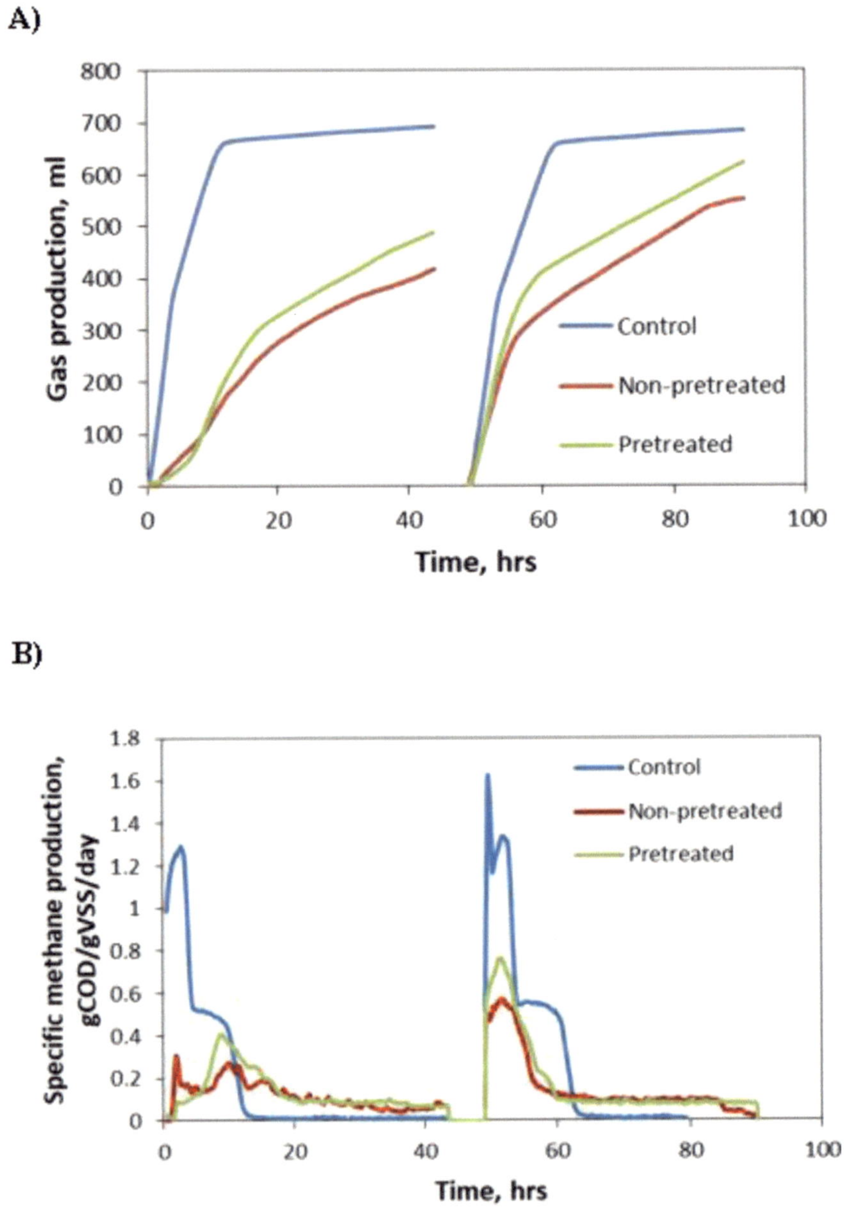

Figure 4.
Total gas production (A) and specific methane production (B) for the distillery wastewater.

3.3.2 Soy protein processing wastewater

Chemical analysis of the soy protein processing wastewater showed that it contains high concentration of sulfates. High sulfate concentration has adverse effect on anaerobic digestion for two reasons: it decreases the content of methane in the biogas, because reduction of sulfur competes with methanogenesis; second, inhibition of methanogenic archaea with hydrogen sulfide can occur [26]. Typically, a safe level of sulfates is considered to be when the ratio of COD to sulfates is at least 10. In our case this ratio is 3–3.5. Thus, the inhibition of anaerobic activity may be expected. Adjustment of pH from original 4 to 6 resulted in the decrease in sulfates concentration by 16.2%. For control, we performed additional removal of sulfates

	Biogas yield, ml		COD reduction, %		Biogas composition, % vol.		
	Cycle 1	Cycle 2	Cycle 1	Cycle 2	CH_4	CO_2	H_2S
Non-pretreated	396 ± 38	548 ± 32	57.5 ± 2.5	80.2 ± 2.7	61.1 ± 1.0	34.8 ± 0.3	1.33 ± 0.02
Pretreated	467 ± 42	611 ± 24	67.8 ± 3.5	89.4 ± 3.2	60.0 ± 2.7	36.5 ± 0.4	1.23 ± 0.03

Table 3.
Biogas yield, % COD reduction, and biogas composition after 40 hours of digestion of the distillery wastewater.

by adding $BaCl_2$. $BaCl_2$ selectively precipitates sulfates by forming insoluble salt $BaSO_4$. As a result of this treatment, 86.4% of sulfates have been removed (sulfates content decreased from 4970 to 600 mg/L).

We performed anaerobic digestion tests of this waste stream using three samples: 1) non-pretreated at initial pH, 2) treated by adjusting pH to 6, and 3) treated with $BaCl_2$ (after pH was adjusted to 6), which is referred to as "w/o sulfates". Results of the test are summarized in **Figure 5** and **Tables 4** and 5. During the first feeding cycle the biogas production from the non-pretreated (pH 4) and pretreated (pH 6) samples is nearly the same. During the second feeding cycle, a significant decrease in the gas production is observed for the non-pretreated sample. The amount of biogas produced after 24 hours from the non-pretreated sample decreased by 40% during the second cycle. The biogas production from the pretreated sample decreased only by 7%. Such a decrease in biogas production can be attributed to the expected inhibition of methanogenic archaea by high sulfates concentration. This assumption is supported by the fact that the sample treated with $BaCl_2$ had higher biogas production than the pretreated sample, and no decrease in the biogas production was observed during the second feeding cycle.

Nearly 60% decrease in methane content during the second feeding cycle was observed in the non-pretreated sample (pH 4). The pretreated sample (pH 6) had lesser (25%) decrease in the methane content during the second cycle. On the other hand, the methane content in the biogas from the sample treated with $BaCl_2$ did not change. These results again suggest the inhibitory effect of the high sulfates concentration.

3.3.3 Oat fiber processing wastewater

Results from the anaerobic digestion test of the oat fiber processing waste stream are summarized in **Figure 6** and **Table 6**. The results show that the pretreatment significantly improves the digestibility of this stream. The amount of biogas produced after 40 hours during the first feeding cycle is 87% higher for the pretreated sample. An increase in digestibility for both samples is observed during the second feeding cycle (53% and 31% for the non-pretreated and pretreated sample, respectively). Upon the pretreatment, COD reduction during the second feeding cycle increased from 48.3% to 77.6%.

The gas quality, however, decreased upon the pretreatment (**Table 6**). The methane content decreased by 20%, carbon dioxide increased by 50%. The reason for this decrease in quality can be high concentration of NaCl, which accumulated as a result of pH adjustment with HCl and NaOH [25, 29].

All three waste streams, especially soy processing wastewater, contain fairly high amount of hydrogen sulfide, which, although unavoidable, is highly undesirable as it decreases the quality of biogas, causes corrosion of the piping, turbines, and other equipment [30]. It also forms a greenhouse gas SO_2 during combustion of H_2S-containing biogas. There is a number of methods to decrease or remove the H_2S

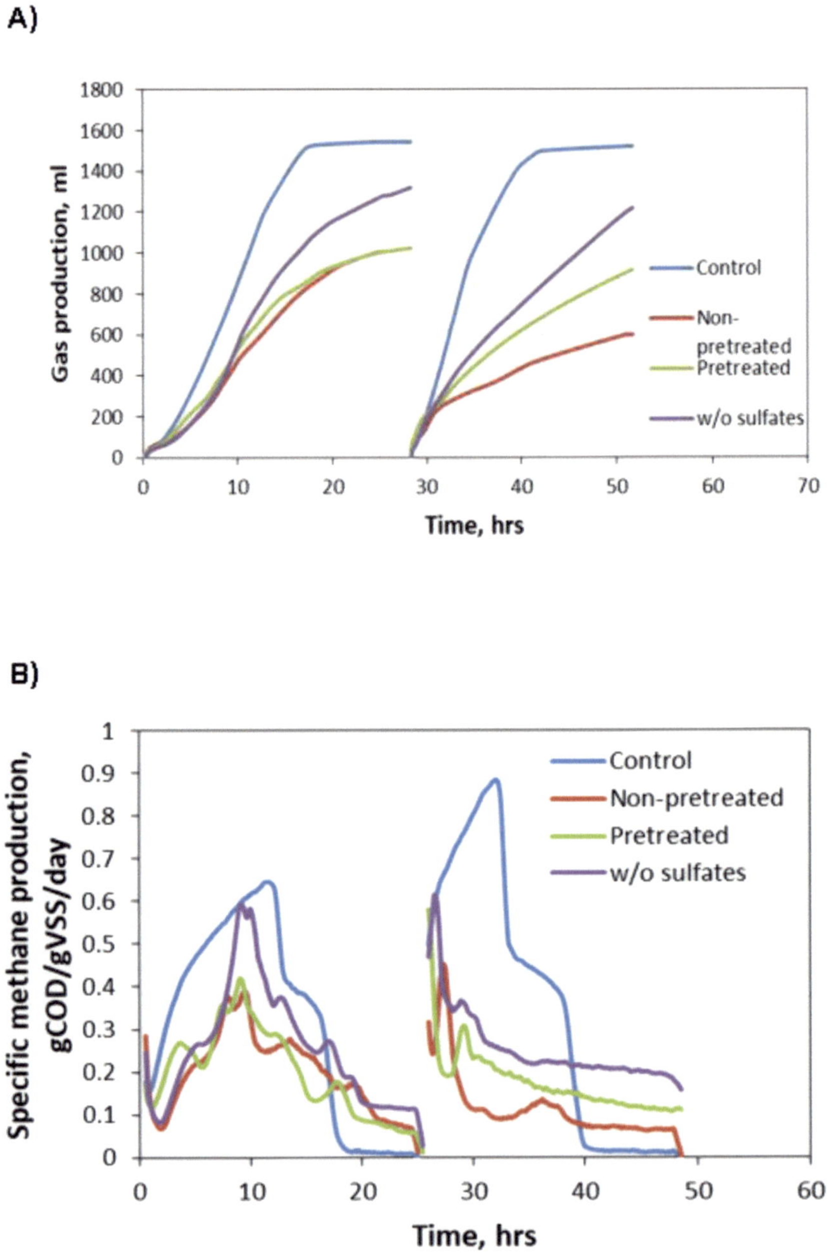

Figure 5.
Total gas production (A) and specific methane production (B) for the soy protein processing wastewater.

	Biogas yield, ml		COD reduction, %	
	Cycle 1	Cycle 2	Cycle 1	Cycle 2
Non-pretreated	992 ± 81	599 ± 62	64.3 ± 4.1	39.3 ± 2.5
Pretreated	990 ± 95	924 ± 85	64.2 ± 4.8	60.7 ± 5.7
W/o sulfates	1247 ± 132	1228 ± 121	80.8 ± 6.4	80.7 ± 7.3

Table 4.
Biogas yield and % COD reduction after 24 hours of digestion for the soy protein processing wastewater.

	Biogas composition, % vol.					
	Cycle 1			Cycle 2		
	CH_4	CO_2	H_2S	CH_4	CO_2	H_2S
Non-pretreated	42.3 ± 0.8	40.1 ± 0.6	3.12 ± 0.08	17.2 ± 0.4	59.3 ± 1.2	3.27 ± 0.05
Pretreated	44.1 ± 0.7	33.8 ± 0.6	2.23 ± 0.07	33.4 ± 0.7	38.4 ± 0.9	3.18 ± 0.07
W/o sulfates	45.3 ± 1.1	34.2 ± 0.8	1.35 ± 0.05	44.1 ± 0.9	45.3 ± 1.7	1.38 ± 0.05

Table 5.
Biogas composition after 24 hours of digestion for the soy protein processing wastewater.

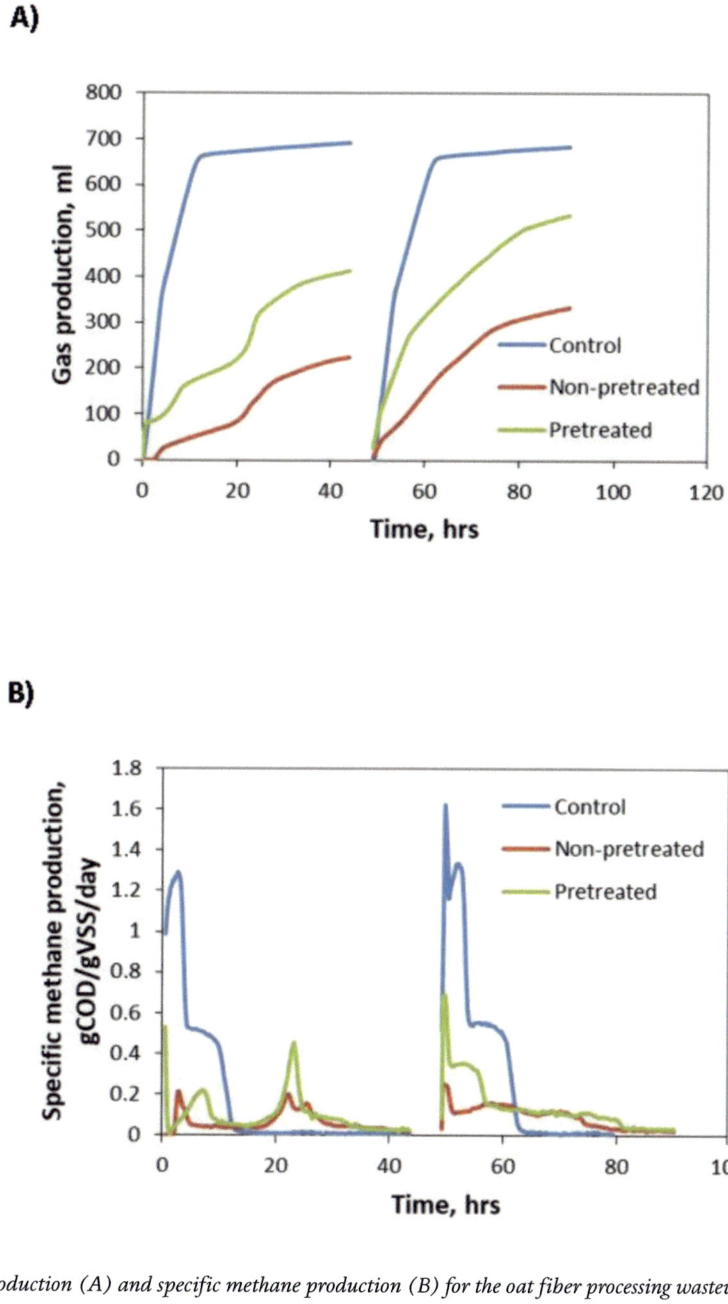

Figure 6.
Total gas production (A) and specific methane production (B) for the oat fiber processing wastewater.

	Biogas yield, ml		COD reduction, %		Biogas composition, % vol		
	Cycle 1	Cycle 2	Cycle 1	Cycle 2	CH_4	CO_2	H_2S
Non-pretreated	216 ± 27	330 ± 43	31.3 ± 4.3	48.3 ± 3.9	65.6 ± 1.7	27.3 ± 0.4	1.24 ± 0.08
Pretreated	404 ± 45	530 ± 61	58.5 ± 5.4	77.6 ± 6.9	52.8 ± 2.1	40.9 ± 0.8	1.45 ± 0.10

Table 6.
Biogas yield, % COD reduction, and biogas composition after 40 hours of digestion for the oat fiber processing wastewater.

content in biogas. They are broadly divided into two categories: 1) post-treatment of biogas and 2) prevention of H_2S formation during the AD process. The first category includes absorption, adsorption, and membrane filtration, and biological filtration techniques [31]. The second category includes in-situ chemical removal and in-situ bioconversion using microaeration [32–34]. Each individual method has its advantages and disadvantages. Therefore, best strategy is integration of several technologies to achieve a balance between efficiency, feasibility, and cost.

3.4 Pilot-scale anaerobic digestion tests

In order to verify that results of batch studies are transferrable on a larger scale, we performed AD tests on a continuous upflow fluidized bed pilot reactor using only one of the tested streams. We selected for this purpose the oat fiber processing wastewater, as it seemed to benefit the most from the pretreatment. Non-pretreated and pretreated wastewater was fed continuously for a 2-week period. The anaerobic sludge in the reactor was preliminary activated by feeding with a standard nutrient solution [24] using ethanol as a source of COD at ~2 g-COD/Lday volumetric loading rate (VLR) for one week. Prior to feeding the wastewater, the biomass in the reactor was starved for 2 days. The COD content of the wastewater was adjusted to 10.0 g/L by dilution with tap water. The wastewater was supplemented with nitrogen in the form of ammonium chloride (10 g per 50 L every second day). COD of influent and effluent, as well as biogas production were measured daily. The results of this test (**Table** 7) indicate that the pretreatment of the wastewater by pH-induced precipitation resulted in the increase of biogas production by 23.1% and increase of the COD removal efficiency by 25.2% compared to the original wastewater. We attribute this improvement to the decrease in the amount of the poorly digestible compounds, such as alkali-soluble polycarbohydrates and lignins, which were precipitated and removed. Methane content, however, was slightly lower in the case of pretreated wastewater, which is consistent with the results of the batch tests. The reason for this is most likely the same as in batch studies – high level of NaCl. Although batch studies did not reveal any adverse effects of this waste stream on the anaerobic biomass, the operation of the pilot reactor was not stable in both non-pretreated and pretreated streams. While the volumetric loading rate (VLR) was kept constant at fairly low level, the volatile fatty acids (VFA) concentration in

	VLR, g-COD/L/day	Gas production, L/day	COD removal efficiency, %	Methane content, % vol.
Non-pretreated	3.0 ± 0.2	65.0 ± 4.0	65.8 ± 3.1	77.8 ± 1.2
Pretreated	3.0 ± 0.2	80.0 ± 3.0	82.4 ± 1.4	74.4 ± 0.9

Table 7.
Summary of anaerobic digestion of the oat fiber processing wastewater using a continuous pilot-scale reactor.

both cases was constantly increasing throughout the entire feeding period, suggesting a possible toxic effect. Elucidation of the long-term effects of the above waste stream on anaerobic biomass was, however, beyond the scope of this study.

4. Conclusions

In this study, pH-induced precipitation has been evaluated as a method of pretreatment of industrial effluents in order to improve anaerobic treatment efficiency. The pH adjustment was done to bring the pH of the solution close to its isoelectric point. Such pretreatment resulted mainly in the removal of suspended and dissolved solids. The effect of the pretreatment was studied on the laboratory and pilot scale using wastewater from local grain processing industries: distillery, soy protein processing, and oat fiber processing plants. The anaerobic digestibility of all three waste streams benefited from the pretreatment. Lab-scale batch AD tests showed the increase in COD reduction from 80.2% to 89.4% for the distillery waste stream, from 39.3% to 60.7% for the wastewater from the soy protein processing, and from 48.3% to 77.6% for the oat fiber processing wastewater. Benefit of the pretreatment was further verified on the pilot scale using an upflow fluidized bed reactor with the oat fiber processing wastewater as a feed. After two weeks of continuous feeding, an increase in the daily biogas production by 23% and COD removal efficiency by 25% has been observed as a result of the pretreatment.

Our lab-scale and pilot-scale AD studies showed a positive effect of the pH-induced precipitation on these waste streams in terms of increased biogas production (11–60%) and COD removal (12–60%), and in some instances reduction in the H_2S content in biogas (8%). This study clearly demonstrated that pH-induced precipitation is an effective pretreatment method to improve AD performance on wastewaters from grain processing industries.

Author details

Jagannadh Satyavolu[1]* and Robert Lupitskyy[2]

1 Conn Center for Renewable Energy Research, University of Louisville, Louisville, KY, USA

2 Ingredion Incorporated, MD, USA

*Address all correspondence to: jagannadh.satyavolu@louisville.edu

References

[1] Mulder, R., Thomas, P., 2003. Biological Wastewater Treatment for Industrial Effluents: Technology and Operation. Paques.

[2] Grasius, M.G., Iyengar, L., Venkobachar, C., 1997. Anaerobic biotechnology for the treatment of wastewaters: a review. J. Sci. Ind. Res. 56, 385-397.

[3] Rajeshwari, K.V., Balakrishnan, M., Kansal, A., Lata, K., Kishore, V.V.N., 2000. State-of-the-art of anaerobic digestion technology for industrial wastewater treatment. Renewable and Sustainable Energy Reviews 4, 135-156.

[4] Van Lier, J.B., Tilche, A., Ahring, B.K., Macarie, H., Moletta, R., Dohanyos, M., Hulshoff, I.W., Lens, P., Verstraete, W., 2001. New perspectives in anaerobic digestion. Water Sci. Technol. 43, 1-18.

[5] Saravanane, R., Murthy, D.V.S., 2000. Application of anaerobic fluidized bed reactors in wastewater treatment: a review. Environ. Manage. Health 11, 97-117.

[6] Heijnen, J., Mulder, A., Enger, W., Hoeks, F., 1989. Review on the application of anaerobic fluidized bed reactors in wastewater treatment. Chem. Eng. J. 41, B37-B50.

[7] Speece, R.E., 1983. Anaerobic biotechnology for industrial wastewater treatment. Environ. Sci. Technol. 17, 416A-427A.

[8] Madhoo, A., 2012. Biogas Production: Pretreatment Methods in Anaerobic Digestion. John Wiley & Sons, Inc. Hoboken, NJ, and Scrivener Publishing LLC, Salem, MA.

[9] Chang, I.-S., Choo, K.-H., Leea, C.-H., Pekb, U.-H., Koh, U.-C., Kim, S.-W., Koh, J.-H., 1994. Application of ceramic membrane as a pretreatment digestion of alcohol-distillery wastes. J. Membrane Sci. 90, 131 – 139.

[10] Sabbah, I., Marsook, T., Basheer, Sobhi., 2004. The effect of pretreatment on anaerobic activity of olive mill wastewater using batch and continuous systems. Process Biochem. 39, 1947-1951.

[11] Mendes, A.A., Pereira, E.B., Furigo Jr., A., de Castro, H.F., 2010. Anaerobic Biodegradability of Dairy Wastewater Pretreated with Porcine Pancreas Lipase. Braz. Arch. Biol. Technol. 53, 1279-1284.

[12] Mobarak-Qamsari, E., Kasra-Kermanshahi, R., Nosrati, M., Amani, T., 2012. Enzymatic Pre-Hydrolysis of high fat Content Dairy Wastewater as a Pretreatment for Anaerobic Digestion. Int. J. Environ. Res. 6, 475-480.

[13] Sigge, G.O., Green, J., du Plessis, K.R., Britz, T.J., 2007. Investigating the Effect of Ozone on the Biodegradability of Distillery Wastewater. S. Afr. J. Enol. Vitic. 28, 155-162.

[14] Bejankiwar R.S., Lokesh K.S., Gowda T.P.H., 2003. Electrochemical pretreatment of wastewater from Color Photograph processing unit. J. Environ. Eng. 129, 1061-1063.

[15] Krishna, B.M., Murthy, U.N., Kumar, B.M., Lokesh, K.S., Savitha, B., 2009. Improvement of BOD_5/COD ratio in pre-treated distillery waste water by electrochemical treatment method. J. Environ. Res. Develop. 4, 90-101.

[16] Rao N.N., Somashekhar K.M., Kaul S.N., Szpyrkowicz L., 1997. Electrochemical oxidation of tannery wastewater. J. Chem. Technol. Biotechnol. 76, 1124-1131.

[17] Lei, X., Sugiura, N., Feng, C., Maekawa, T. 2007. Pretreatment of

anaerobic digestion effluent with ammonia stripping and biogas purification. J. Hazard. Mater. 145, 391-397.

[18] Shao, L., Wang, X., Xu, H., He, P. 2012. Enhanced anaerobic digestion and sludge dewaterability by alkaline pretreatment and its mechanism. J. Environ. Sci. (China) 24, 1731-1738.

[19] Alqaralleh, R.M. 2012. Effect of Alkaline Pretreatment on Anaerobic Digestion of Organic Fraction of Municipal Solid Waste. Thesis. University of Ottawa, Ottawa, Canada.

[20] Bramaud, C., Aimar, P., Daufin, G., 1997. Whey Protein Fractionation: Isoelectric Precipitation of α-Lactalbumin under Gentle Heat Treatment. Biotechnol. Bioeng. 56, 391-397.

[21] Hoflanda, G.W., de Rijkea, A., Thieringc, R., van der Wielen, L.A.M., Witkamp, G.-J., 2000. Isoelectric precipitation of soybean protein using carbon dioxide as a volatile acid. J. Chromatogr. B 743, 357 –368.

[22] Cui, R., Jahng, D. 2006. Enhanced methane production from anaerobic digestion of disintegrated and deproteinized excess sludge. Biotechnol. Lett. 28, 531-538.

[23] Delgenès JP, Penaud V, Torrijos M, Moletta R. 2000. Investigations on the changes in anaerobic biodegradability and biotoxicity of an industrial microbial biomass induced by a thermochemical pretreatment. Water Sci Technol. 41, 137-144.

[24] Young, J.C., Cowan, R.M., 2004. Respirometry for Environmental Science and Engineering. SJ Enterprises, Springdale.

[25] Chen, Y., Cheng, J.J., Creamer, K.S. 2008. Inhibition of anaerobic digestion process: A review. Biores. Technol. 99, 4044-4064.

[26] Maillacheruvu, K.Y., Parkin, G.F., Peng, C.Y., Kuo, W.C., Oonge, Z.I., Lebduschka, V., 1993. Sulfide toxicity in anaerobic systems fed sulfate and various organics. Water Environment Research. 65, 100-109.

[27] Straka, F., Jenicek, P., Zabranska, J., Dohanyos, M., Kuncarova, M. 2007. Anaerobic fermentation of biomass and wastes with respect to sulfur and nitrogen contents in treated materials. Proceedings Sardinia 2007, Eleventh International Waste Management and Landfill Symposium. S. Margherita di Pula, Cagliari, Italy.

[28] Peter, C.M., Han, Y., Boling-Frankenbach, S.D., Parsons, C.M., Baker, D.H. 2000. Limiting order of amino acids and the effects of phytase on protein quality in corn gluten meal fed to young chicks. J. Anim. Sci. 78, 2150-2156.

[29] Hierholtzer, A., Akunna, J.C., 2012. Modelling sodium inhibition on the anaerobic digestion process. Water Sci Technol. 66, 1565-1573.

[30] Charles, W., Cord-Ruwisch, R., Costa, G.H.M., Spencer, P., 2006. Solutions to a combined problem of excessive hydrogen sulfide in biogas and struvite scaling. Water Science and Technology, 53, 203-210.

[31] Zulkefli, N.N., Jahim, J., Masdar, M.S., Majlan, E.H., 2016. Overview of H2S Removal Technologies from Biogas Production. International Journal of Applied Engineering Research 11, 10060-10066.

[32] Ge, H.Q., Zhang, L., Batstone, Keller, J., 2013. Impact of Iron Salt Dosage to Sewers on Downstream Anaerobic Sludge Digesters: Sulfide Control and Methane Production. Journal of Environmental Engineering-Asce, 139, 594-601.

[33] Lupitskyy, R., Alvarez-Fonseca, D., Herde, Z.D., Satyavolu, J., 2018. In-situ

prevention of hydrogen sulfide formation during anaerobic digestion using zinc oxide nanowires. J. Env. Chem. Eng. 6, 110-118.

[34] Okoro, O.V., Sun, Z., 2019. Desulphurisation osf Biogas: A Systematic Qualitative and Economic-Based Quantitative Review of Alternative Strategies. Chem Engineering 3, 76.

Chapter 3

Employment of Organic Residues for Methane Production: The Use of Wastes of the Pulp and Paper Industry to Produce Biogas - A Case Study

Alfredo de Jesús Martínez-Roldán,
María Dolores Josefina Rodríguez-Rosales,
Sergio Valle Cervantes and Thania Azucena Mendez-Perez

Abstract

Many organic residues are being wasted since they are not given a comprehensive management; anaerobic digestion is an alternative to reduce the impact of these residues, and to produce biogas. The chapter includes the state of art about biogas and energy production, and later, the analysis of a study case focusing on the use of pulp and paper wastes to produce biogas. The study was carried out through anaerobic digestion at a bench scale using three temperature phases to treat primary and secondary sludge, establishing operational parameters such as temperature, retention time, and organic loadings. Monitoring of volume, methane concentration in the biogas, volatile solids reduction, volatile fatty acids during the process, the performance of the process in function of methane produced per volatile solids removed is calculated. This case study shows that it is feasible to use the sludge from the company's wastewater treatment plant (WWTP) for the generation of biogas, thus reducing waste management problems.

Keywords: anaerobic digestion, biogas, bioenergy, biosolids, paper industry

1. Introduction

The indiscriminate use of raw materials and fossil fuels has led to an infinity of environmental problems, such as water reservoir pollution, acidification of the oceans, loss of ecosystem diversity, and a concentration increase in certain gaseous pollutants in the atmosphere [1]. To reduce dependence on oil and decrease the CO_2 concentration to revert the climate change, it is necessary to use renewable energy sources [1, 2]. Among the possible renewable energy sources biogas, stands out especially, when biogas is obtaining from waste produced from different productive activities [3]. The biogas generation through anaerobic digestion generates several environmental benefits, such as reducing greenhouse gas emissions, depletion in the residuals environmental impact, the clean energy generation, and the possibility of using the generated biosolid as a soil improver or fertilizer, among others [4].

1.1 Anaerobic digestion of waste

Under anaerobic conditions decomposition of matter produces a gaseous mixture known as biogas. Methane is the main fuel gas in the biogas mixture, and to be used as a fuel, its content must be above 45% of the total composition of biogas [5]. Biogas general characteristics are listed in **Table 1**.

Biomasses such as food industry waste, animal excreta, straws, residual planta and municipal waste under anaerobic digestion are able to produce biogas [4, 8, 9]. Through several biochemical steps the macromolecules of organic matter are transforming into CH_4, CO_2 and H_2S under anaerobic digestion [8–11]. However, the organic matter characteristics must allow being used as an energy source for a set of microorganisms that will make the digestion process possible. Therefore, not only a supply of the main nutrients (carbon and nitrogen) is necessary, but also a balance of micro and macro nutrients [12]. Carbon and nitrogen are the principal sources of food for methanogenic microorganisms, and the proportion between these nutrients must be adequate for the correct operation of the process. It is known that the approximate proportion of carbon (C) and nitrogen (N) consumption by bacteria is 30:1 (C/N), this being the optimum point. On the other hand, if there is a ratio of 35:1 the process is inhibited due to a lack of nitrogen, and if it is 8:1 the inhibition occurs due to the formation of ammonia [4, 13].

Anaerobic biodegradation of complex organic materials is a multi-stage process where solid materials are first hydrolyzed, polysaccharides to sugars and alcohols, proteins to polypeptides and amino acids, lipids to long-chain fatty acids (LCFA), and glycerol. From these, the fermentative bacteria produce short-chain fatty acids (SCFA), hydrogen (H_2), and carbon dioxide (CO_2), and ammonia producing by the fermentation of amino acids. Subsequently, acetogenic bacteria from non-acetic FA and neutral materials such as ethanol produce H_2, acetate, and CO_2, which are used by methanogenic bacteria to produce CH_4, CO_2, and H_2O [10, 12, 14]. The process can be divided into four steps according to the proposed models and the complex inter-microbial relationships that carry it out.

Composition	55–70% Methane (CH_4) 30–45% Carbon dioxide (CO_2) 0–10% Nitrogen (N_2) 0–1% Hydrogen (H_2) 0–2% Oxygen (O_2) 0–3% Hydrogen sulfide (H_2S) <0,5 mg/m^3 Siloxanes
Heat content	6.0–6.5 [kWh m^{-3}]
Fuel equivalent	0,60–0,65 [L petroleum m^{-3} biogas]
Explosion limit	6–12% biogas in the air
Ignition temperature	650–750°C (with the mentioned CH_4 content)
Critical pressure	74–88 [atm]
Critical temperature	−82,5 [°C]
Normal density	1,2 [kg m^{-3}]

Table 1.
Common characteristics of biogas [6, 7].

In the specific case of using waste sludge from Wastewater Treatment Plants (WWTP), the nutrients are in the necessary proportions and concentrations. However, the sludge generated in the plants of the forestry or paper industry contains high concentrations of cellulose, and this unbalances the C/N ratio [15]. Besides, lignin can cause toxicity problems and decrease the efficiency of the anaerobic digestion process [10, 15, 16].

In the hydrolysis process, macromolecules such as proteins, lipids, carbohydrates, and nucleic acids are transformed into oligomers (fatty acids, carbohydrates, amino acids, nitrogenous bases, and aromatic compounds) [17, 18]. The bacteria involved in the process are a very complex mix of many genera, most of which are obligate anaerobes; however, some facultative anaerobic bacteria such as streptococci and other enteric microorganisms may be present. This type of microorganisms ferments a great variety of complex organic molecules such as polysaccharides, lipids, and proteins, turning them into a wide range of end products such as acetic acid, a mixture of H_2 and CO_2, mono carbon compounds, organic acids with more than two carbon atoms, and compounds such as propanol, and butanol [10, 19]. The optimum pH for hydrolysis varies according to the substrate. For easily degradable carbohydrates, hydrolysis proceeds in an accelerated manner at pH between 5.5 and 6.5 [17].

In acidogenesis, the pH value decreases, going from 7.0 to values around 5.0; in this stage, the bacteria ferment the soluble products of hydrolysis, mainly hydrogen and volatile fatty acids, and long-chain fatty acids also produce acetate or propionate by ß-oxidation. Thus, together, hydrolytic and acidogenic bacteria convert complex substrates to precursors of methanogenesis: H_2, CO_2, and acetate, in addition to AGV and other reduced compounds, ethanol, lactate [9, 10].

In the acetogenesis stage, organisms that favor an acidic environment participate; during this stage, volatile fatty acids and nitrogenous compounds are slowly transformed. During this stage, the pH value increases from values around 5.0 to values around 6.8. The metabolic products of acetogenic bacteria are converted into substrates for methanogens by the activity of the acetogenic bacteria constituting the third level or trophic group in the population sequence that occurs in anaerobic digestion. The metabolic result of this group is the formation of acetate, H_2, and CO_2. These bacteria are known as hydrogen obligate acetogenic bacteria. This trophic group must have a symbiotic relationship with hydrogenophilic archaea, since they consume the hydrogen produced by the former, thus avoiding its inhibition by-product accumulation [17, 19].

In the last digestion stage, known as methanogenesis, the volatile fatty acid content drops to less than 500 ppm. The pH value increases from 6.8 to 7.4, producing large volumes of gases with 65 to 70% CH_4, around 30% CO_2, and other inert gases such as N_2. Methanogenic archaea are responsible for producing methane from various substrates, with acetate being responsible for approximately 73% of the methane produced. Methanogenic archaea are strict anaerobes, very sensitive to oxygen as they require negative oxidoreductive potentials lower than −50 mV to grow [20]. The main products of this type of treatment are biogas and biosolids, which are used as a source of energy and as a fertilizer respectively. An additional benefit of this type of processing is that a load of pathogenic organisms in the sludge is very low, as is the mass of the sludge. The main uses of sludge from bioreactors are soil conditioning, use as fertilizer, and use for the generation of vegetation cover in sanitary landfills or for the recovery of degraded soils or sites, and also in their bioremediation [17].

1.2 Factors involved in anaerobic digestion

Biomass has a varied composition that includes different organic and inorganic compounds. To optimize the anaerobic digestion process and biogas production, parameters such as chemical composition, operational parameters such as temperature, pH, loading rate, alkalinity, biodegradability, bioaccessibility, bioavailability, and the initial characterization of substrates [11, 21].

1.2.1 Temperature

Temperature is one of the principal survival factors of microorganisms during the anaerobic digestion process [10]. The management of the temperature range is useful to differentiate the type digestion processes. Three operating ranges can be used in an anaerobic digester: psychrophilic (~ 25° C), mesophilic (~ 35° C), and thermophilic (~ 55° C). Microorganisms grow best in temperature ranges between 35 and 55° C. An increase in temperature has a positive effect on the metabolic rate and accelerates the degradation of biomass; however, the use of a thermophilic range is difficult to control and generates energy consumption to maintain the constant temperature of the reactor. In general, the mesophilic process often involves a diversity of microorganisms and is more stable than the thermophilic process. Temperature is one of the principal parameters for microorganisms to grow, degrade organic matter, and consequently, biogas to be produced [11, 21].

1.2.2 pH

The pH value is one of the main operational factors that can affect the anaerobic digestion process. That is because most of the microorganisms prefer a neutral pH range. In the biogas production process, some organisms require a different growth pH. However, the most favorable pH range to obtain maximum biogas production is 6.8 to 7.2. In the anaerobic digestion process, methanogenic microorganisms are too sensitive to pH variations and prefer a pH of around 7.0 [11, 22].

Acidogenic microorganisms are less sensitive to pH and are tolerable in the 4.0–8.5 range. However, the optimal pH for hydrolysis and acidogenesis is between 5.5 and 6.5 [11, 22]. The pH value is an important factor because it influences the ratio of ionized and non-ionized forms. This is because excessive hydrogen, sulfur, fatty acids, and ammonia are toxic in their non-ionized forms. Generally, the pH value indicates a healthy environment for the digester microorganisms [11, 22, 23].

1.2.3 Alkalinity

Alkalinity is the ability of a system to maintain a certain pH. It is a measure of the buffer capacity of the system. The higher the alkalinity, the better the pH despite an increase in H+ generation. In systems where anaerobic digestion is performed, the buffer system is due to the presence of carbonates, in particular the presence of the bicarbonate ion HCO_3^-. Since acidogenic bacteria have a higher activity than methanogenic bacteria, they are capable of causing acidification in the reactor, in case of organic matter overloads. This acidification can be avoided by maintaining an optimal buffer capacity in the digester. Alkalinity is useful for buffering purposes, at typical operating pH values [21, 22].

1.2.4 Volatile fatty acids

The concentration of volatile fatty acids (VFA) product of the fermentation has great importance in the anaerobic digestion process. This because the VFA can acidify the reactor, causing the failure of the process. Under normal operating conditions, the concentration of VFA in the effluent must be very low or negligible, less than 100 mg L^{-1}. On the contrary, if there is a high concentration, it can cause inhibition of methane-forming archaea. The VFA/alkalinity ratio is also an indicator of stability. A ratio greater than 0.4 indicates an immediate failure [21, 24].

1.2.5 Chemical composition of substrates

Substrates chemical composition characterization is useful to identify the appropriate substrates to carry out the anaerobic digestions. Substrates contain the full range of simple and complex chemical compounds, and the proportion of them will depend on their sources (agricultural agriculture and animal manure, municipal, food, and industrial waste). Specific organic compounds may predominate. Although, most of the time the exact composition of the substrates is difficult to determine [22, 24].

1.3 Temperature regimes in anaerobic digestion

As commented in a previous section, the temperature regime is important when looking for the conditions that allow increasing the degradation of organic matter and the production of biomass. For this reason, each of the possible regimes will be briefly analyzed.

1.3.1 Mesophilic anaerobic digestion

It is the type of conventional anaerobic digestion carried out in a temperature regime ranging from 33 to 35° C, which can have a system that allows mixing of the sludge. In this configuration, the retention times are usually long, VS reduction reaches around 40 to 48%. It presents a problem of foam generation, and destruction of the pathogens is not carried out. The quality of the biogas in this type of digestion is good, however, the volumes generated are not so considerable, which in terms of profitability makes it inefficient [25, 26]. It has been founding that for retention times between 5 and 55 days, the methane concentration can be between 62 and 66%, and the reduction in volatile solids can reach 32 to 40% for retention times between 15 and 30 days [27].

1.3.2 Thermophilic anaerobic digestion

The waste sludge treatment process in thermophilic terms is one of the most studied at present. This type of process, carried at a temperature of 50 to 55° C, allows an improvement in the deployment of retention times, and destruction of pathogens. Popat et al. (2010) report that the reduction of most pathogens can occur between 13 and 15 days at constant temperatures between 51 and 55°C. However, the energy cost resulting from the treatment puts it into consideration [28]. The VS reduction percentages are around 50 to 60%, which makes it a point of study for its improvement in energy terms [26, 29]. Besides, Wahidunnabi & Eskicioglu (2014) and Yu et al. (2014) reported that VS removal efficiencies for thermophilic systems range from 40 to 50%. Regarding the production of biogas, with values around 0.30 m^3 CH_4 (kg of VS fed)$^{-1}$ [26, 30].

1.3.3 Three-phase temperature anaerobic digestion

This digestion is a combination of acid/gas phases and temperature phases, from which a good removal of volatile solids is obtaining, it does not produce fetid odors, and the retention times are shorter [29, 30]. Riau, de la Rubia, & Pérez (2010) carried out a configuration for this type of digestion, where the phases are delimiting by time and temperature. The mesophilic from 1 to 3 days, the thermophilic from 5 to 15 days, and a mesophilic with a retention time from 5 to 15 days; The results of his research were 55% SV reductions, coliform and pathogen reduction, as well as a volumetric gas production of ~5.5 L_{CH4} (kgVS fed)$^{-1}$ [31]. Similarly, the experiments carried out by Kim, Novak, & Higgins (2011), affirm the effectiveness of the combination of three temperature phases. In their results, they obtained a VS reduction of about 57% [32].

1.4 Sewage sludge and its use to produce biogas

Most conventional wastewater treatment systems generate large amounts of waste products, which are called sludge. The composition and quantity of the sludge depend on the raw wastewater characteristics and the wastewater treatment process. The main constituents of wastewater disposed of in treatment plants include garbage, sand, foam, and sludge. The sludge extracted and produced in wastewater treatment operations and processes is generally a liquid or a liquid-semi-solid with a high solids content between 0.25–12% [33, 34]. The different treatments to process sludge vary according to the source and type of wastewater from which they are deriving, the process used to treat the wastewater, and the final disposal of the sludge. Sludge is by far the constituent with the highest volume removed in wastewater treatment, so its treatment and disposal are probably the most complex problem [34].

The biological wastewater treatment process produces different types of sludge within each of the individual processes, such as (1) primary sludge produced during the primary wastewater treatment processes; this occurs after sieving and de-sanding. The composition of the sludge depends on the characteristics of the wastewater. It mainly contains large undissolved solids that generally carry on a large amount of organic material, vegetable matter, paper, and other materials. (2) Activated sludge coming from the removal of dissolved organic matter during aerobic or anaerobic treatment of wastewater. This sludge is generally in the form of flocs that contain living and dead biomass. (3) Tertiary sludge, which is produced through subsequent treatment processes, with the addition of flocculating agents [35]. The processes for treating sludge vary according to the type of wastewater from which they are deriving, the process used to treat them, and the last disposal method to which the sludge will be destining. The sludge treatment main objectives are to reduce mass and volume, to handling it easily and to increase its biological stability in order to produce a sufficiently harmless material for its disposal [35, 36].

1.5 Biogas in México

Energy is a vital supply for the development of any society, but when talking about energy, it encompasses aspects such as use and abuse, source of supply, pollution generated in its generation, danger to society in cases of accidents, etc. Global energy consumption has doubled in the last 25 years. Estimation for the next 25 years shows that there will be an increase of 70%. In developing countries, the above will be reflected mainly due to globalization, population growth, and economic growth. Besides, the consumption of fossil fuels is no longer sustainable due to its early depletion, the increase in its price, and the damage it has caused to the environment [37]. México has an enormous potential in renewable resources, and thanks to the reforms implemented in the energy sector, barriers that impede the development of new projects and clean technologies have been eliminated, achieving increases in a clean generation far above fossil energy. According to the clean energy progress report, from 2016 to 2017, fossil generation grew by 2.07% and clean by 6.98% [38].

In México, the production of electrical energy is based mainly on the consumption of different fossil fuels, reaching more than 90% of the total, highlighting the use of oil, natural gas, and coal; on the other hand, the fraction of energy obtained by renewable means is 7.5%, and biogas only contributes 0.02% [39, 40]. However, it is important to note that the percentage covered by renewable energies increases every year, although without yet becoming one of the most important sources [39]. However, the National Energy Strategy aims for approximately 35% of the country's consumed energy to be renewable origin by 2024 and marks that 50% of the consumed energy in 2050 be clean [41]. Experts estimate that the generation of biogas from waste has great potential in México, specifically for the use of livestock waste. It is estimated that from anaerobic digestion of them, little more than 100 million cubic meters of biogas could be generated per year, which would allow covering little more than 8% of the national energy demand [42, 43]. On the other hand, in the case of the wastewater treatment plants, the potential is slightly lower, reaching projections of around 75 million cubic meters per year for 2024; however, studies on wastewater treatment plants of the industrial sector are still needed since their effluents and operating conditions are specific, making it difficult to generalize about possible production values and biogas yields [42].

1.6 Waste from the paper industry to produce biogas

The pulp and paper industry produces large amounts of highly polluting waste; this cause the wastewater and consequently the treatment plant sludge to have particular characteristics [44]. It is estimated that up to 1 m3 of residual sludge can be generated per ton of paper produced, which will contain between 45 and 55% organic matter in addition to the presence of other pollutants and COD between 4,000 and 15,000 ppm depending on whether it is primary or secondary sludge [44, 45]. Due to its high content of organic matter, it can be used for biogas generation from anaerobic digestion, being able to achieve high values of biogas production as well as high conversion efficiencies [45].

The primary sludge is producing when clarifying the wastewater from the process. This sludge has a high content of lignocellulosic material; the fiber content is variable depending on the type of process, and the dewatering of this type of sludge is relatively simple [44, 46]. The solids content can be up to 48%, while the volatile solids and total organic carbon can reach values of 33 and 19%; the presence of heavy metals such as chromium, zinc, nickel, among others stands out [46]. On the other hand, secondary sludge is the sludge generated when carrying out the biological treatment (aerobic, anaerobic, activated sludge) of the wastewater generated in the process. The secondary sludge is recovered in the clarification phase from the treated water and is normally mixed with the primary sludge to incinerate it or to deposit it in a landfill [44, 46]. Traditionally, the sludge generated in the pulp and paper industry is mixed (primary and secondary), later they are dried, and finally, they are used as fuel when incinerated, another alternative is to place them in landfills. However, the large amount of organic matter causes its weathering to generate a great amount of greenhouse gases, so this strategy is not currently allowed in many countries. [46–48]. Since the majority fraction of the industry and paper sludge is organic matter, its use in an anaerobic digestion process has been proposed to recover energy from them. [46]. However, the process is not very efficient because a large part of the organic matter is composed of cellulose and lignin, for which various authors propose the use of pretreatment strategies that allow the breaking of the fibers and increase the efficiency of the process of anaerobic digestion [46, 49, 50]. However, the sludge characteristics depend on the operating conditions, the raw material, among others therefore the anaerobic digestion process must be adjusted and specifically designed.

2. Case of study: the use of wastes of the pulp and paper industry to produce biogas a case of study

This study was carried out to treat residual sludge from a paper-producing industry. A company and leader in the manufacture of paper and cardboard packaging, which treats a flow of 80 to 100 L s^{-1} of wastewater, which results in annual production of primary and secondary sludge of 5,400 to 6,000, and of 4,300 to 5,000 tons $yr.^{-1}$, of primary and secondary sludge, respectively.

This papermaking company uses recycled paper as raw material to manufacture paper with three quality grades: linerboard paper for corrugated packaging, medium paper for corrugated packaging, and white top paper for corrugated packaging. That generates a variation in the wastewater characteristics resulting from the process, making it difficult for the company to treat activated sludge. This wastewater treatment consists of screening and desander pretreatment, primary clarification of primary treatment, and biological treatment. The solids from the primary settler and the flotation process are mixed and concentrated through a sludge press. The effluent from the primary clarification is neutralized and transported to the activated sludge treatment. The mixed liquor flows from the reactor to the secondary settler, where the produced sludge and the clarified effluent are separated. Primary and secondary sludge do not receive any treatment and are disposing on the land of the company. For all the anterior, this study of anaerobic digestions is the first step taken to research giving added value to the generated sludge and avoiding contamination in soils and phreatic levels.

The use of these residual sludge for the generation of biogas was studied through anaerobic digestion, using three bioreactors, one operating at mesophilic temperature (M), another at thermophilic temperature (T), and another at three temperature phases (mesophilic, thermophilic, and mesophilic) (M-T-M).

2.1 Methodologies

Primary and secondary sludge were sampling in the industry. The primary sludge was taking before the sludge press, and the secondary sludge from the sludge return line to the oxidation lagoon. Samples were transporting to the laboratory for their characterization. A mixture of primary and secondary sludge was preparing in a 50:50 ratio, thickening and concentrating the sludge to prepare an organic loading of 1.4 kg m^{-3} d^{-1}.

Total solids (TS) and volatile solids (VS), pH, alkalinity, total nitrogen, volatile acids, chemical oxygen demand (COD) and total and fecal coliforms were measured according to the Standard Methods [39].

Elemental composition (C, H, N, S) and protein were conducted according to the procedure ISO-16948: 2015 [40].

Gas production was measured by displacement of an acidified brine solution (NaCl and H_2SO_4) in graduated cylinders. +.

Volatile fatty acids (VFA) was reassured by titration according to [41].

Biogas composition by a LandTec® gas analyzer.

2.2 Biodigester operation

Three stainless steel bioreactors (14-L each) were used to carry out the experimental anaerobic digestion process. The bioreactors had inlet and outlet valves for feeding and collecting biogas. Also, bioreactors had mechanical stainless-steel propeller-type stirrers, driven by an Arrow brand motor, model 350. The shakers were programmed to shake the content for three minutes every twenty minutes to keep the sample homogeneous by shaking the reactors 20 times per day for 3 minutes the intervals between each shaking were 20 minutes. The digesters were providing with submersible electrical resistance and temperature control. The bioreactors were operating with an organic load mixture of 1.4 kg m-3 d-1 of primary and secondary sludge, in a ratio of 50:50 and a retention time of 30 days. One reactor was operating at a mesophilic temperature (M) of 35°C, another at a termophilic temperature (T) of 55°C, and the other at three temperature phases (M-T-M), mesophilic 35°C, thermophilic 55°C, and mesophilic 35°C. The reactors were operating in semi-batch mode, feeding, and removing substrate every third, day and performing the analysis of Total and volatile solids, pH, alkalinity and acidity, volatile acids, total Kjeldhal nitrogen, total coliforms, fecal coliforms and measuring the volume and biogas composition.

3. Results and discussion

3.1 Initial characterization of residual sludge

Table 2 shows the results obtained from the physicochemical and biological characterization for the primary sludge, secondary sludge, and the 50:50 mixture. The percentage of total solids is within the range of 5 to 9% according to [35]. The analysis were carried out in triplicate for each of the parameters analyzed. The cellulose present in the primary and secondary sludge is the result of the fact that its recovery is not total during the flotation process, and there is a great loss of these residues and has the potential to be reused for obtaining energy due to their high

Parameter	Primary sludge	Secondary sludge	Mixture 50:50
Total solids (mg L^{-1})	81655	78310	92380
Volatile solids (mg L^{-1})	43900	33225	42565
Total solids (%)	5.54	7.01	9.20
Volatile solids (%)	53.76	42.20	46.08
pH	6.07	6.75	6.10
Alkalinity (mg_{CaCO3} L^{-1})	1245	465	777
C (%)	36.35	46.42	48.68
H (%)	33.66	37.41	41.33
N (%)	1.27	3.67	5.98
S (%)	0	0	0
Protein (%)	7.97	23.43	10.98
Kjeldhal total nitrogen (mg L^{-1})	434	590	896
DQO (mg L^{-1})	235	560	550
Total coliforms (NPM gST^{-1})	5.69E10^8	2.20E10^9	1.31E10^8
Fecal coliforms (NPM gST^{-1})	4.7E10^7	3.20E10^8	2.40E10^7
Celulose (%)	92.68	92.65	92.65
Heat energy Kcal Kg^{-1}	81655	2501.16	1044.36

Table 2.
Physicochemical and biological characterization of the primary, secondary, and mixed sludge.

calorific content. The cellulose concentration in the secondary sludge is due to the low biodegradability of its biological wastewater treatment [42]. The results of the alkalinity in the sludge are determined to give sludge buffer capacity, because that the anaerobic digestion process needs to withstand the changes in pH as the process progresses [43]. The content of carbon, hydrogen, nitrogen, total nitrogen, and proteins are necessary substrates for the reproduction of microorganisms and the generation of biogas [22]. The concentration of coliforms present in the sludge exceeds the Official Mexican Standard NOM-004-SEMARNAT-2002, so they require treatment for their disposal [44].

The results obtained for the 50% sludge mixture (primary sludge and secondary sludge) indicate that the combination of both substrates maintains conditions of total solids, volatile pH to carry out anaerobic digestions, having 42.5 g L^{-1} which, corresponds to 46% of SV of organic matter to be degraded, contained in the mixture of substrates.

3.2 Volatile solids

Figure 1 shows the VS results for the 50:50 ratio of substrates with organic load (OL) of 1.4 Kg m^{-3} d^{-1} with a retention time of 30 days. It can be seen that, during the digestions in the three treatments, there was the removal of solids, the final removals of SV (%) for each treatment in its temperature phase were M = 52.49, T = 57.76, and M-T-M = 58.61.

3.3 pH, alkalinity y total volatile acids

Figure 2A and **B** show the behavior of pH and alkalinity parameters, respectively, during anaerobic digestion. **Figure 2A** shows that the T and M-T-M bioreactors managed to increase their pH to 7.3. The opposite case occurred with the mesophilic bioreactor where the increase of pH was only 6.7. **Figure 3B** shows that in bioreactor M there was a variation in alkalinity due to the low pH obtained values. For the T and M-T-M bioreactors, there was a decrease in alkalinity, reaching concentrations of 900 mg L^{-1} after 20 days.

Figure 3 presents the VFA concentration, which decreased during the digestion process. During the digestions, there was no accumulation of VFA therefore, the process was not destabilized. It shows that the concentrations of VFA were decreasing throughout the process in the three bioreactors. Starting with 7100, 7800, and

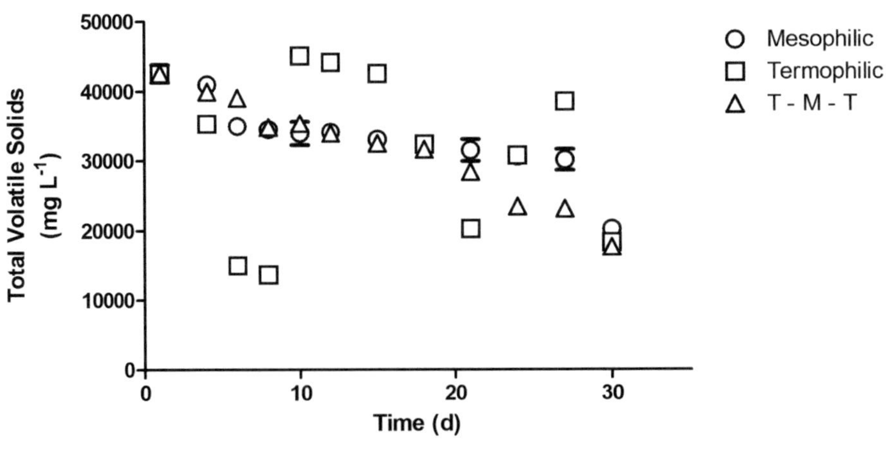

Figure 1.
Volatile solids concentration during anaerobic digestion processes.

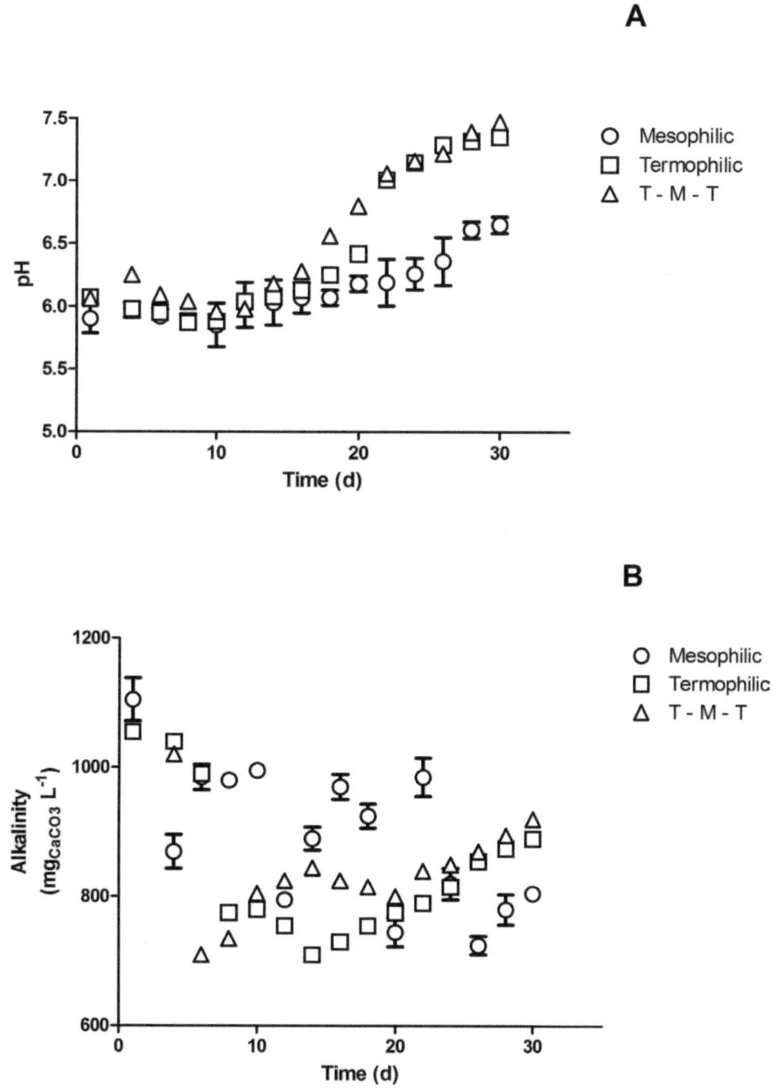

Figure 2.
Behavior of pH (A) and alkalinity (B) during anaerobic digestion processes.

840 mg L^{-1} for the M, T, and M-T-M bioreactors, respectively. At the end of the treatments, the concentrations of 540, 640, 610 mg L^{-1} for the M, T, and M-T-M bioreactors, respectively. The buffer capacity in the digesters, neutralized the possible accumulation of volatile acids and maintained the pH values to stabilize the anaerobic digestion.

3.4 Organic and ammonia nitrogen

Figure 4 shows the results of ammonia nitrogen during the experimentation, and it is observing how the ammonia nitrogen increased through the process for the three different temperatures. The increase in the concentration of ammonia nitrogen was not inhibitory for the development of the digestion process because all the bioreactors at the different temperatures presented biogas production.

Figure 5A and **B** shows the behavior of the biogas volume and methane fraction. It is showing that the T and M-T-M reactors generated a greater volume of biogas

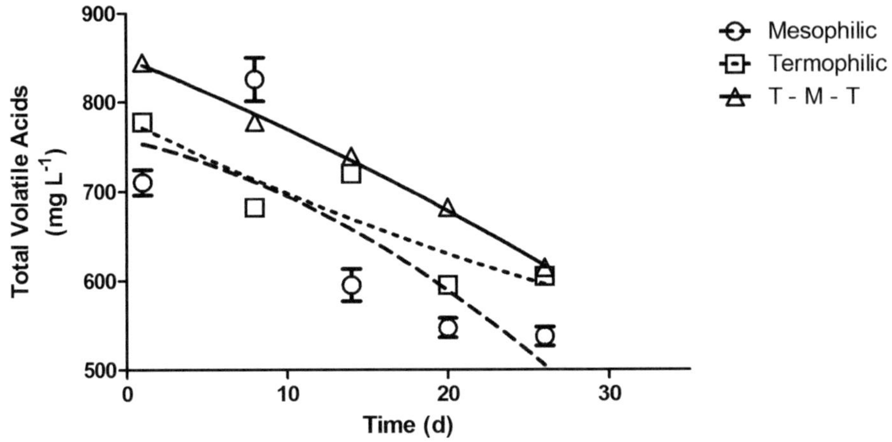

Figure 3.
Concentration of total volatile fatty acids during anaerobic digestion processes.

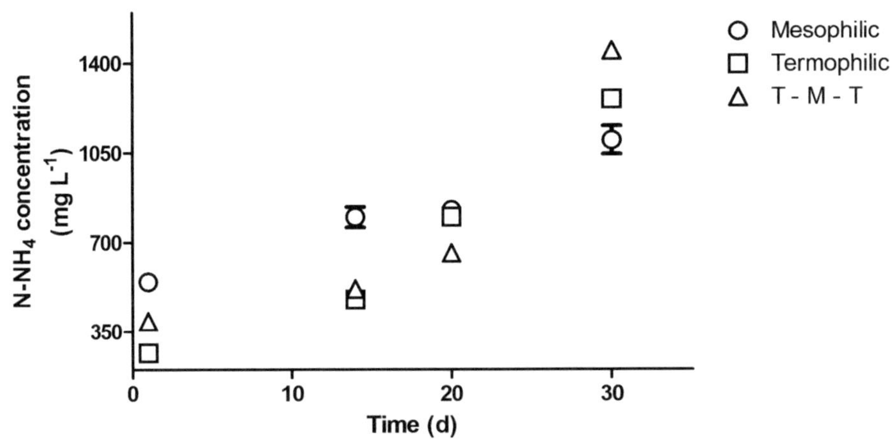

Figure 4.
Concentration of N-NH$_4$ during anaerobic digestion processes.

than the M bioreactor, which presented too low biogas volumes. The methane fraction was higher in the M-T-M bioreactor where a value above 60% was obtained.

Figure 6 presents the methane yields in the anaerobic digestion processes. The M-T-M bioreactor resulted in a higher methane yield until day 24, after this time there was a decrease in methane yield. Methane yield was very low for the M and T bioreactors because of the conditions, but for the M bioreactor the yield was the lowest.

According to the literature review, there are research studies on different biomasses that can be processed in anaerobic digestions, such as agro-industrial, livestock, forestry residues, sludge from sewage treatment plants, industrial residues, where the biogas and methane yields are reporting when digesting these substrates. However, for particular wastes from industry using recycled paper raw materials, there are no studies to date. There are studies of the pulp and paper industry where other types of pollutants are generated from the chemical process, a case that does not apply to this industry. There is research on anaerobic digestions, always seeking to obtain high methane yields for reuse as biofuel or energy production. There is

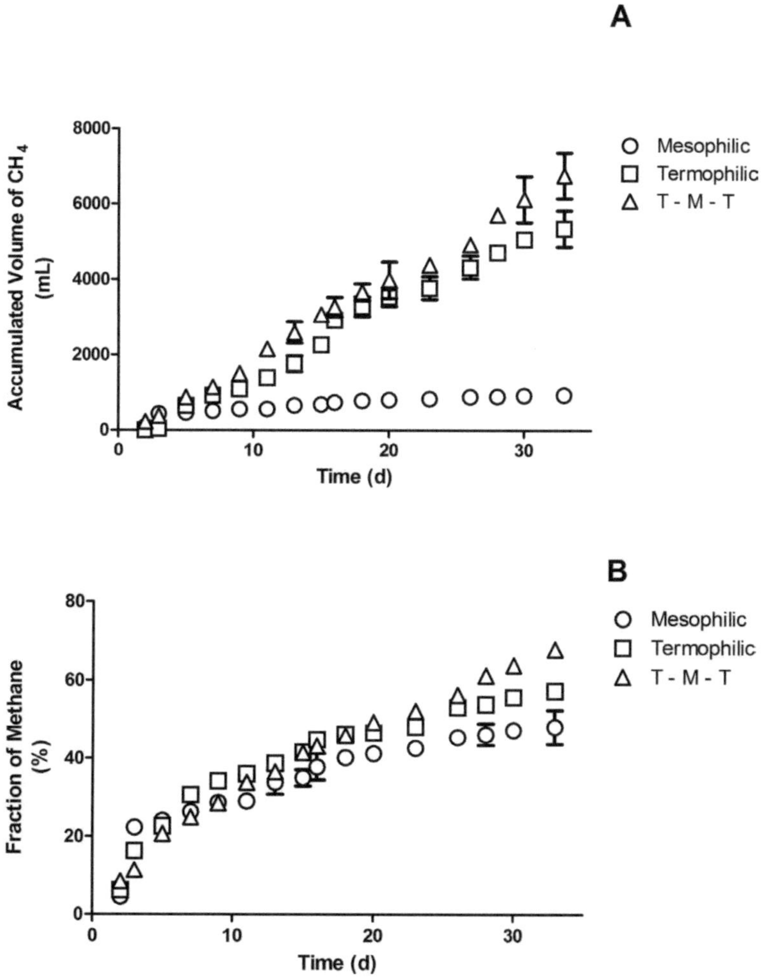

Figure 5.
Behavior of the biogas volume (A) and methane fraction (B) during the anaerobic digestions.

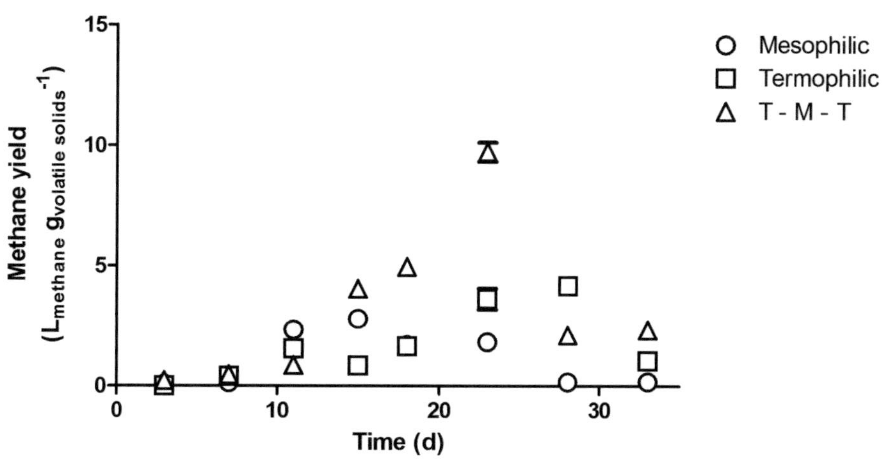

Figure 6.
Methane yields in the anaerobic digestion processes.

also too much difference in the investigations carried out on AD in the production of methane or biogas because it is not only the substrates or cosubstrates but also the influence on the operating conditions of the bioreactors such as temperature, time retention, organic loads, pH, and C / N ratio, among others. The scope in AD is to optimize the process with the best conditions that result in high methane yields. For this reason, this study serves as the basis for future research to work on the combination of lignocellulosic waste from a primary treatment in combination with secondary sludge from an activated sludge process, both resulting from the treatment of wastewater from the industry under study. During the anaerobic digestions, the mixture into the bioreactors was agitated because one of the AD main problems is the mixing of the substrates into the bioreactor. Bad mixing results in low methane production due to the non-homogenization of the substrates [45].

The reduction of volatile solids was observed in the three treatments reaching yields greater than 50%, being greater in the digestion of the three stages M-T-M. The biodegradation of the organic compounds depends on the substrates to digesting.

During the anaerobic digestions, the pH remained between 5.8 and 6.5 in the first 18 days later after day 20 the thermophilic and three-stage MTM digestions the pH increased above 7, which did not happen with the mesophilic digestion that reached a pH of 6.6. The variation of the buffer capacity was influenced so that pH in the mesophilic anaerobic digestion did not increase its pH beyond 6.6. At pH lower than 6.6, the growth rate of methanogens was reduced and the activity of archaea-methanogenic bacteria is reduced both at low and high pH's [46]. Even so, in anaerobic digestion, M was produced in biogas, with 45% methane. For the other T and M-T-M digestions above 50 and 60% methane were obtained, respectively.

The samples in the three anaerobic digestions showed ammonia concentrations lower than 5000 mg L^{-1}, which represented avoiding the inhibition of the VFA during the digestions [47]. It is observing that during the digestions the VFA decreased during the process however, in the M digestion the production of VFA was lower.

The temperature influences VFA production. It has been reporting that at thermophilic temperatures, VFA yields are higher due to faster acclimatization and more active acidogenesis, than at mesophilic temperatures [48].

However, there are other authors [49] that at thermophilic temperatures of 45 to 70°C it does not affect the production of VFA, finding controversies and inconsistencies in research due to the difference between microbial species, raw materials or substrates to be digested. Likewise, the use of different methodologies in AD affects the methane yield in equivalent substrates, making their comparison difficult [50–52]. The thermophilic process presents a better performance at the beginning than the mesophilic digestion due to the accelerated process of hydrolysis [53]. Higher methane yields are produced in the T and M-T-M, the latter having an advantage over the thermophilic since more than 60% of methane was obtained, which is considered biogas rich in methane [54]. Fecal coliform analyzes were performed during anaerobic digestions to determine their stabilization. Thermophilic digestion is a proven technology to produce class "A" biosolids, NOM-004-SEMARNAT-002. Where it turned out that the T and M-T-M digestions manage to obtain a biosolid with fecal coliforms lower than the norm.

4. Conclusions

The sludge generated from the paper process contains a high content of cellulose, which can be used by some microorganisms present in the secondary sludge. These microorganisms could be used as a potential raw material for the production of methane [15, 42].

Anaerobic digestion of the primary and secondary sludge showed promising results for methane production. The research carried out with a mixture of primary and secondary sludge is to increase the yield of biogas and methane, since each one of the substrates provides different physicochemical and biological characteristics. The primary sludge calorific value is high compared to the secondary sludge, and mixing both sludge benefited the anaerobic digestion process.

A higher methane yield was obtained in the digestion of three M-T-M phases with a value of 24.75 L of methane (gr of VS)$^{-1}$, also, a higher volume and percentage of methane, with values of 7000 mL and 67%, respectively.

The three-phase M-T-M process started with a pH value of 6.2 and was increased through digestion, reaching a pH of 7.6. The alkalinity was kept between 800 and 900 mg L^{-1}, making the digestion process tolerate changes during the anaerobic digestion phases. That allowed no accumulation of organic acids, which diminish the production of methane gas [55].

The reduction of volatile solids occurred in the three digestions, with the thermophilic phase presenting a larger removal with 52%, followed by the three phases with 47%, and finally the mesophilic with 30%.

It was found that thermophilic and three-phase digestion have advantages over mesophilic digestion related to the destruction of bacteria and pathogens [56] in this study. The thermophilic and three-phase digestion stabilized the sludge by destroying bacteria since in the thermophilic process and the three-phase M-T-M process, fecal coliforms were eliminated on days 15 and 12, respectively, classifying these sludge as Class A according to the official Mexican stàndard NOM 004-SEMARNAT.

Acknowledgements

The authors would like to thank the TecNM/IT Durango for their help and financial support.

Conflict of interest

The authors declare no conflict of interest.

Author details

Alfredo de Jesús Martínez-Roldán[1,2], María Dolores Josefina Rodríguez-Rosales[2]*, Sergio Valle Cervantes[2] and Thania Azucena Mendez-Perez[2]

1 CONACyT-TecNM/TecNM/IT de Durango, Master on Environmental Systems, Durango, Dgo, México

2 TecNM/IT de Durango, Master in Environmental Systems, Durango, Dgo, México

*Address all correspondence to: mdjoserr@itdurango.edu.mx

References

[1] Cavicchioli R, Ripple WJ, Timmis KN, Azam F, Bakken LR, Baylis M, et al. Scientists' warning to humanity: microorganisms and climate change. Nature Reviews Microbiology. 2019;17:569-586. https://doi.org/10.1038/s41579-019-0222-5

[2] Bai X, Dawson RJ, Ürge-Vorsatz D, Delgado GC, Salisu Barau A, Dhakal S, et al. Six research priorities for cities and climate change. Nature. 2018;555:23-25. https://doi.org/10.1038/d41586-018-02409-z

[3] Gålfalk M, Olofsson G, Crill P, Bastviken D. Making methane visible. Nature Climate Change. 2016;6(4):426-430. https://doi.org/10.1038/nclimate2877

[4] Panigrahi S, Dubey BK. A critical review on operating parameters and strategies to improve the biogas yield from anaerobic digestion of organic fraction of municipal solid waste. Renewable Energy. 2019;143:779-797. https://doi.org/10.1016/j.renene.2019.05.040

[5] Hanjie Z. Sludge treatment to increase biogas production. 1st ed. Trita-LWR Degree Project; 2010. 32 pp. https://www.sjostadsverket.se/download/18.79cc091012c369366d9800017089/1350483757500/LWR_EX_10_20.pdf

[6] Rasi S. Biogas composition and upgrading to biomethane University of Jyväskylä. 2009. http://urn.fi/URN:ISBN:978-951-39-3618-1

[7] Bruijstens AJ, Kadijk G, Bleuanus S. Biogas composition and engine performance, including database and biogas property model. Stockholm: Biogasmax. https://www.researchgate.net/profile/Robert_Cloudt/publication/239579361_Biogas_Composition_and_Engine_Performance_Including_Database_and_Biogas_Property_Model/links/57208be508aead26e71ca416/Biogas-Composition-and-Engine-Performance-Including-Database-and-Biogas-Property-Model.pdf

[8] Alatriste-Mondragón F, Samar P, Cox HHJ, Ahring BK, Iranpour R. Anaerobic Codigestion of municipal, farm, and industrial organic wastes: a survey of recent literature. Water Environment Research. 2006;78(6):607-636. https://doi.org/10.2175/106143006X111673

[9] Achinas S, Achinas V, Euverink GJW. A technological overview of biogas production from biowaste. Engineering. 2017;3(3):299-307. https://doi.org/10.1016/J.ENG.2017.03.002

[10] Li Y, Chen Y, Wu J. Enhancement of methane production in anaerobic digestion process: A review. Applied Energy. 2019;240:120-137. https://doi.org/10.1016/j.apenergy.2019.01.243

[11] Schnürer A, Jarvis Å. Microbiological handbook for biogas plants. Swedish Waste Management U2009:03. Swedish Gas Centre Report. 2010 http://www.emrg.it/Lezioni_Energia_Rinnovabile/Microbiological_handbook_for_biogas.pdf

[12] Yu Z, Schanbacher FL. Production of methane biogas as fuel through anaerobic digestion. In: Singh OV and Harvey SP editors. Sustainable Biotechnology: Sources of renewable energy. Springer Netherlands; 2010 p. 105-27. https://doi.org/10.1007/978-90-481-3295-9_6

[13] Zahan Z, Othman MZ, Muster TH. Anaerobic digestion/co-digestion kinetic potentials of different agro-industrial wastes: A comparative batch study for C/N optimisation. Waste Management. 2018;71:663-674. https://doi.org/10.1016/j.wasman.2017.08.014

[14] Duran M, Tepe N, Yurtsever D, Punzi VL, Bruno C, Mehta RJ. Bioaugmenting anaerobic digestion of biosolids with selected strains of *Bacillus*, *Pseudomonas*, and *Actinomycetes* species for increased methanogenesis and odor control. Applied Microbiology and Biotechnology. 2006;73(4):960-966. https://doi.org/10.1007/s00253-006-0548-6

[15] Li W, Khalid H, Zhu Z, Zhang R, Liu G, Chen C, et al. Methane production through anaerobic digestion: Participation and digestion characteristics of cellulose, hemicellulose and lignin. Applied Energy. 2018;226:1219-28. https://doi.org/10.1016/j.apenergy.2018.05.055

[16] Vashi H, Iorhemen OT, Tay JH. Degradation of industrial tannin and lignin from pulp mill effluent by aerobic granular sludge technology. Journal of Water Process Engineering. 2018;26:38-45. https://doi.org/10.1016/j.jwpe.2018.09.002

[17] Rittmann BE. Opportunities for renewable bioenergy using microorganisms Biotechnology and Bioengineering. 2008;100:203-212. https://doi.org/10.1002/bit.21875

[18] Rittmann BE, McCarty, PL. Methanogenesis. In Rittmann BE, McCarty PL, editors. Environmental Biotechnology: Principles and Applications. 2nd Ed. McGraw-Hill Education; 2020. https://www.accessengineeringlibrary.com/content/book/9781260441604/chapter/chapter10

[19] Rittmann BE, McCarty PL. Microbial Ecology. In Rittmann BE, McCarty PL, editors. Environmental Biotechnology: Principles and Applications. 2nd Ed. McGraw-Hill Education; 2020. https://www.accessengineeringlibrary.com/content/book/9781260441604/chapter/chapter4

[20] Ganesh SR, Kumar G, Banu R, Xia A, Periyasamy S, Dattatraya Saratale GA critical review on anaerobic digestion of microalgae and macroalgae and co-digestion of biomass for enhanced methane generation. Bioresource Technology. 2018;262:319-332. https://doi.org/10.1016/j.biortech.2018.03.030

[21] Escamilla-Alvarado C, Ríos-Leal E, Ponce-Noyola MT, Poggi-Varaldo HM. Gas biofuels from solid substrate hydrogenogenic-methanogenic fermentation of the organic fraction of municipal solid waste. Process Biochemistry. 2012;47:1572-1587. https://doi.org/10.1016/j.procbio.2011.12.006

[22] van Lier JB, Mahmoud N, Zeeman G. Anaerobic wastewater treatment. In: Guang-Hao C, van Loosdrecht MCM, Ekama GA, Brdjanovic D. Biological Wastewater Treatment: Principles, Modeling and Design. IWA Publishing; 2020. p. 701-756. https://doi.org/10.2166/9781789060362_0701

[23] Kallistova AY, Goel G, Nozhevnikova AN. Microbial diversity of methanogenic communities in the systems for anaerobic treatment of organic waste. Microbiology. 2014;83:462-483. https://doi.org/10.1134/S0026261714050142

[24] Dirnena I, Dimanta I, Gruduls A, Kleperis J, Elferts D, Nikolajeva V. Influence of the initial acidification step on biogas production and composition. Biotechnology and Applied Biochemistry. 2021;61(3):316-321 https://doi.org/10.1002/bab.1163

[25] Ganidi N, Tyrrel S, Cartmell E. The effect of organic loading rate on foam initiation during mesophilic anaerobic digestion of municipal wastewater sludge. Bioresourse Technology. 2011;102(12):6637-43. https://doi.org/10.1016/j.biortech.2011.03.057

[26] Yu D, Kurola JM, Lähde K, Kymäläinen M, Sinkkonen A, Romantschuk M. Biogas production and methanogenic archaeal community in mesophilic and thermophilic anaerobic co-digestion processes. Journal of Environmental Management. 2014;143:54-50 . https://doi.org/10.1016/j.jenvman.2014.04.025

[27] Hwang K-Y, Shin E-B, Choi H-B. A mechanical pretreatment of waste activated sludge for improvement of anaerobic digestion system. Water Science and Technology. 1997;36(12):111-6. https://doi.org/10.1016/S0273-1223(97)00731-2

[28] Popat SC, Yates M V., Deshusses MA. Kinetics of inactivation of indicator pathogens during thermophilic anaerobic digestion. Water Research. 2010;44(20):5965-5972. https://doi.org/10.1016/j.watres.2010.07.045

[29] Willis J, Schafer P. Advances in thermophilic anaerobic digestion. Proceedings of the Water Environment Federation. 2006;7:5378-5392.

[30] Wahidunnabi AK, Eskicioglu C. High pressure homogenization and two-phased anaerobic digestion for enhanced biogas conversion from municipal waste sludge. Water Research. 2014;66:430-446. https://doi.org/10.1016/j.watres.2014.08.045

[31] Riau V, de la Rubia MÁ, Pérez M. Temperature-phased anaerobic digestion (TPAD) to obtain Class A biosolids. A discontinuous study. Bioresourse Technology. 2010;101(1):65-70. https://doi.org/10.1016/j.biortech.2009.07.072

[32] Kim J, Novak J, Higgins M. Multistaged anaerobic sludge digestion processes. Journal of Environmental Engineering. 2011;137(8):746-753. https://doi.org/10.1061/(ASCE)EE.1943-7870.0000372

[33] Bhattacharya SK, Madura RL, Walling DA, Farrell JB. Volatile solids reduction in two-phase and conventional anaerobic sludge digestion. Water Research. 1996;30(5):1041-1048. https://doi.org/10.1016/0043-1354(95)00252-9

[34] Daud MK, Rizvi H, Akram MF, Ali S, Rizwan M, Nafees M, et al. Review of upflow anaerobic sludge blanket reactor technology: Effect of different parameters and developments for domestic wastewater treatment. Journal of Chemistry. 2018. https://doi.org/10.1155/2018/1596319

[35] Metcalf & Eddy I, Tchobanoglous G, Burton F, Stensel H. Wastewater Engineering: Treatment and Reuse. 4th ed. McGraw-Hill, Inc.; 2002. 1848 p.

[36] Praspaliauskas M, Pedišius N. A review of sludge characteristics in Lithuania's wastewater treatment plants and perspectives of its usage in thermal processes. Renewable and Sustainable Energy Reviews. 2017;67:899-907.

[37] Mundial Bank. Consumo de energía eléctrica (kWh per cápita) | Data [Internet].. Available from: https://datos.bancomundial.org/indicator/EG.USE.ELEC.KH.PC [Accessed 2021 Feb 06]

[38] Abergel T, Brown A, Cazzola P, Dockweiler S, Dulac J, et al. Energy technology perspectives 2017 - Catalysing Energy Technology Transformations. OECD; 2017, 443 p.

[39] Baird RB, Rice EW, Eaton AD. Standard methods for the examination of water and wastewater.23th ed. APHA, AWWA, WEF. 2017. p. 1496.

[40] ISO 16948:2015 - Solid biofuels - Determination of total content of carbon, hydrogen and nitrogen. Available from: https://www.iso.org/standard/58004.html [Accessed 2021 Feb 06].

[41] Le X, Huang Y, Ren Y, Chen L. Anaerobic digestion characteristics and microbial structure of agricultural wastes under gradient organic loadings. Transactions of the Chinese Society of Agricultural Engineering. 2017;33(17):239-247.

[42] Gonzalez-Estrella J, Asato CM, Jerke AC, Stone JJ, Gilcrease PC. Effect of structural carbohydrates and lignin content on the anaerobic digestion of paper and paper board materials by anaerobic granular sludge. Biotechnology and Bioengineering. 2017;114(5):951-960. http://doi.wiley.com/10.1002/bit.26228

[43] Kolbl S, Forte-Tavčer P, Stres B. Potential for valorization of dehydrated paper pulp sludge for biogas production: Addition of selected hydrolytic enzymes in semi-continuous anaerobic digestion assays. Energy. 2017;126:326-334. https://doi.org/10.1016/j.energy.2017.03.050

[44] Norma Oficial Mexicana NOM-004-SEMARNAT-2002, Protección ambiental.- Lodos y biosólidos.- Especificaciones y límites máximos permisibles de contaminantes para su aprovechamiento y disposición final - 15 de Agosto de 2003 - DOF. Diario Oficial de la Federación Available from: https://biblioteca.semarnat.gob.mx/janium-bin/detalle.pl?Id=20210206132626

[45] Amani T, Nosrati M, Sreekrishnan TR. Anaerobic digestion from the viewpoint of microbiological, chemical, and operational aspects - A review. Environmental Reviews. 2010;18:255-78. https://doi.org/10.1139/A10-011

[46] Zhang P, Chen Y, Zhou Q. Waste activated sludge hydrolysis and short-chain fatty acids accumulation under mesophilic and thermophilic conditions: Effect of pH. Water Research. 2009;43(15):3735-42. https://doi.org/10.1016/j.watres.2009.05.036

[47] Salehizadeh H, Van Loosdrecht MCM. Production of polyhydroxyalkanoates by mixed culture: Recent trends and biotechnological importance. Biotechnology Advances. 2004;22(3):261-79. https://doi.org/10.1016/j.biotechadv.2003.09.003

[48] Mengmeng C, Hong C, Qingliang Z, Shirley SN, Jie R. Optimal production of polyhydroxyalkanoates (PHA) in activated sludge fed by volatile fatty acids (VFAs) generated from alkaline excess sludge fermentation. Bioresourse Technology. 2009;100(3):1399-405. https://doi.org/10.1016/j.biortech.2008.09.014

[49] Yu J, Zheng M, Tao T, Zuo J, Wang K. Waste activated sludge treatment based on temperature staged and biologically phased anaerobic digestion system. Journal of Environmental Sciences. 2013;25(10):2056-64. https://doi.org/10.1016/S1001-0742(12)60266-6

[50] Angelidaki I, Alves M, Bolzonella D, Borzacconi L, Campos JL, Guwy AJ, et al. Defining the biomethane potential (BMP) of solid organic wastes and energy crops: A proposed protocol for batch assays. Water Science & Technology. 2009;59(5):927-34. https://doi.org/10.2166/wst.2009.040

[51] Holliger C, Alves M, Andrade D, Angelidaki I, Astals S, Baier U, et al. Towards a standardization of biomethane potential tests. Water Science & Technology. 2016;74(11):2515-22. https://doi.org/10.2166/wst.2016.336

[52] Raposo F, Fernández-Cegrí V, De la Rubia MA, Borja R, Béline F, Cavinato C, et al. Biochemical methane potential (BMP) of solid organic substrates: evaluation of anaerobic biodegradability using data from an international interlaboratory study. Journal of Chemical Technology and

Biotechnology. 2011;86(8):1088-98. https://doi.org/10.1002/jctb.2622

[53] Lin L, Yang L, Li Y. Effect of feedstock components on thermophilic solid-state anaerobic digestion of yard trimmings. Energy and Fuels. 2015;29(6):3699-706. https://doi.org/10.1021/acs.energyfuels.5b00301

[54] Abbasi SA, Nipaney PC. Generation of biogas from *Salvinia molesta* (Mitchell) on a commercial biogas digester. Environmental Technology Letters. 1984;5(1-11):75-80. https://doi.org/10.1080/09593338409384255

[55] Kroeker EJ, Schulte DD, Sparling AB, Lapp HM. Anaerobic Treatment Process Stability. Journal (Water Pollution Control Federation). 1979;51:718-727. https://www.jstor.org/stable/25039893?seq=1

[56] Ghosh S, Ombregt JP, Pipyn P. Methane production from industrial wastes by two-phase anaerobic digestion. Water Research. 1985;19(9):1083-1088. https://doi.org/10.1016/0043-1354(85)90343-4

Section 2

Processing of Household Wastes

Chapter 4

Small-Size Biogas Technology Applications for Rural Areas in the Context of Developing Countries

Martina Pilloni and Tareq Abu Hamed

Abstract

The world's rural population surpasses the three billion people mainly located in Africa and Asia; roughly half the global population lives in the countryside. Access to modern fuels is a challenge for rural people compared to their urban counterparts, which can easily access infrastructures and commercial energy. In developing countries rural populations commonly depend on traditional biomass for cooking and heating. A key strategy in tackling the energy needs of those rural populations is to advance their energy ladder from the inefficient, traditional domestic burn of biomass, organic waste, and animal manure. Governments and non-governmental institutions have supported small biogas digesters in rural areas, mainly in Asia, South America, and Africa, over the last 50 years. This chapter reviews the literature to offer an overview of experimental and theoretical evidence regarding the characteristics of design, construction material, feedstock, and operation parameters that made anaerobic digestion in small digesters a valuable source. Small-scale rural biogas digesters can generate environmental, health, and social benefits to rural areas with a net positive impact on energy access. Remarkable improvement in living standards was achieved with small inputs of the methane, produced via anaerobic digestion; however, challenges associated with lack of technical skills, awareness, and education remain and obstruct biogas' full potential in rural areas, mainly in developing countries.

Keywords: small-scale biogas installations, household biodigesters, rural livelihood, biogas in developing countries, energy access

1. Introduction

Anaerobic digestion is a technology that converts waste into energy. The produced biogas is considered as the primary energy output. The percentage of methane in the biogas is responsible for its calorific value, which is generally considered high [1]. Biogas can substitute oil, coal, and natural gas. Biogas can also be upgraded and directly used in natural gas pipelines and vehicles. The exploitation of fossil fuels and natural resources has increased greenhouse gas (GHG) emissions, deforestation, infertility of land, consumption, and water pollution. Biogas as a source of energy may help to mitigate those problems and reduce global warming. Moreover, using anaerobic fermentation to convert organic waste into fuel has many advantages over the use of crops to generate biofuels: it limits land use, food scarcity, and biodiversity damage. Thus, biogas represents an ethical choice for energy production [2]. In terms of net energy generation, the methane from anaerobic digestion is considered competitive regarding efficiency and costs compared to other biomass energies [3], and it is better from an ecological point of view [4].

Those benefits are already attributed to anaerobic digestion and biogas technology worldwide; however, the contribution of small-scale biogas installations to rural areas in developing countries has a wealthier meaning, and this chapter is aimed to disclose and discuss such value.

The design of biogas technology varies depending on the country, climatic conditions, and the feedstock availability; moreover, it depends on the policy regulations such as waste and energy programs and energy accessibility and affordability. Thus, biogas production may vary from different ranging set-ups, from backyard systems to large industrial plants. In developing countries, the domestic small-scale biogas installations, also called household anaerobic digesters, are the most diffused systems in the rural areas [5]. Those systems volume generally ranges up to 10 m^3 [6]. The digester size is limited by the available feedstock volume originated by the household and easily accessible; the most common feedstocks are manure from animal husbandries, food waste, small-agriculture waste, and sewage sludge. The household systems represent an effective strategy to enhance rural household life quality because it simultaneously advances sanitation and rural ecology and increases energy availability and incomes from the small agricultural activities [7]. The most common energy use of household biogas is for cooking and lighting [8]. Those systems have been successfully employed worldwide with governments and institutions' involvement, supporting household biogas' diffusion throughout subsidy schemes and programs of planning, design, building, and maintenance [9].

The chapter aims to offer an overview to the whole scientific community, to those already interested in biogas technologies but not expressly focused on developing countries and those who started to face the topic. It seems essential to attract new interest in biogas technology from practitioners involved in energy poverty and sustainable development for the Global South, the chapter is also directed to them.

2. Methodology

An overall evaluation of recent literature is used to compare relevant cases that disclose theoretical and practical assessments of small-scale biogas installations in rural areas. The literature review included only publications focused on developing economies; thus, papers were selected to achieve insights on the recent and current status of small-size biogas installations in such contexts. The information gathered is summarized here as principal aspects, designs, materials, and operations as they are applied to the most diffused small-scale and household installations in rural areas. Moreover, the literature data are compared to extract and discuss the relevance that small-biogas technology has for impoverished communities and the prevailing barriers that still slow down, or even prevent, biogas technology diffusion.

3. Rural areas in developing countries: defining the context

The world's rural population has been growing slowly since 1950. There are 3.4 billion people who live in rural areas around the world, 90% of them live in Africa and Asia. India (893 million) and China (578 million) represent 43% of the world's rural population. As the rural population worldwide became more sedentary and grew in population and density, the related environmental and public health problems increased. The population growth determined an increase of consumption needs, and several effects are due to such increased demands. The more prevailing demand is the need for food that can be met through intensification and extensification of agricultural land use; these two responses to the increased food demand are often led by the lack of technological innovation and efficient practices. Indeed, if the land is available, the land extensification is more likely to happen; depending on geographical area, communities may cut trees in lowland forest, use highland slopes in high mountainous regions, or root out brushes in semi-arid zones. Thus, in the absence of environmental controls and adequate rural policies, as generally occurred in the past, the consequences have been deforestation, soil degradation, and desertification in areas already marked by poverty. The population growth determines an increase in energy demand for cooking and heating. In developing countries fuelwood is the cheapest and primary source of energy for cooking and heating. If fuelwood is available in the vicinity, local deforestation results, otherwise deforestation occurs elsewhere also at a long distance from the community [10]. Besides deforestation, which represents an urgent issue in the current climate change era [7], fuelwood's use creates other concerns that need attention. In terms of environmental concern, the diffused utilization of inefficient biomass source contributes to the greenhouse gas emissions [11]. Indeed, biomass as wood and charcoal, both used in poor rural areas, is not sustainable, and when it is partly burnt, it causes emissions that contribute to global warming [12]. As a health concern, because of the use of wood stoves by the rural households, a high level of exposure to Respirable Suspended Particulate Matter (RSPM) from the fuelwood stoves smoke generates health hazards mainly for women and children [13]. From the perspective of social-economic aspects, the women and children are the main fuelwood gatherers (even from long distance), and the fuelwood is collected at the expense of their labor, time, and drudgery [14], and it withdraws them from opportunities of education and incomes.

In developing countries, the rural areas suffer more than urban clusters from lack of basic infrastructure with low access rates to clean water, household sanitation [15], and waste management [16], which determine high public health risk, which is exacerbated by the continuous growth of population and density. The absence of such infrastructures drives rural communities toward practices that negatively affect their surrounding with contamination and pollution of land, water, and air due to unmanaged organic waste from the household and livestock [17, 18]. Practices of burning organic waste as animal dung and crop residues represent how rural communities meet their cooking and heating needs, although it is inefficient and detrimental for the health [19].

Rural areas also suffer from the limited or absent electricity supply and distribution infrastructures, so rural populations have low access to electricity. It was estimated that 770 million people in 2019 were without electricity access; in Africa in the year 2020 there were 592 milion people without electricity access, and the Sub-Saharan represents the region where the access deficit is higher [20]. Such a struggle in energy access drives rural populations to rely on traditional biomass resources or become dependent on imported fossil fuel derivates. However, as already described, these resources have negative impacts on health and the environment and weaken those economies which are already fragile [21].

4. Developing countries: small-scale biogas programs for rural areas

The attention to small-scale biogas technologies has increased in the last decades globally, with fast development and diffusion in rural areas in Asia, Africa, and Latin America [6]. The mass dissemination was dependent on central government programs and long-term political support [22]. Between 1970 and 1985, China established a program for promoting and facilitating the installation of biogas in every rural household; the program brought the installation of 4.7 million household digesters by the end of 1988 [23]. A further increase was observed starting from the end of the 20th century, China registered more than 26 million biogas household installations in 2007 [5], and 43 million biogas users were counted in 2013 [24]. Since 1981, India had the National Project on Biogas Development (NPBD) with various training and development programs and financial support [25]. As a result of Governments' subsidies, over five million household biodigesters were installed in 2014 [26]. In Latin America, the introduction of biogas technologies for households was driven by the energy crisis in the 1970s when the Latin American Energy Commission (OLADE) prompted installations in several counties.

Moreover, the network Biodigesters in Latin America and the Caribbean (RedBioLAC) were created in 2009 to promote household, community, and farm-scale digesters in Latin America [27]. Bolivia stands out among the Countries involved in the network, with over 1000 domestic biogas digesters installed in 2014 [28]. Many other small scale biogas programs were implemented for developing rural areas [19, 29]. In Africa, over 44% increase in domestic digesters installed between 2011 and 2012, and about 60,000 digesters were in Burkina Faso, Ethiopia, Kenya, Tanzania, and Uganda in 2015 [30].

In many other cases, the success of biogas implementation was due to the combination of governmental support and non-profit organizations. Netherland Development Organization (SNV), based in Netherland, had supported national biogas programs impacting more than 2.9 million people in different continents [31].

5. Biogas production and potential in developing countries

The biogas energy supply is a valuable sector for the bioenergy industry. In 2017, 1.33 EJ of biogas was produced globally, representing 2% of the total biomass produced for energy purposes, but it has the potential to develop much more. Europe leads in biogas supply for more than 50% of the global supply, Asia follows it with 31%, and America with 14% [32].

Although the developing countries displayed more barriers for biogas application, some countries such as China [33], South Africa [34], Ghana, Rwanda, and Tanzania [35] produce biogas from large scale institutional plants using similar technology implemented in developed countries.

However, in developing countries, biogas is predominantly produced on a small and domestic scale. In China, the 43 million small-scale biogas installations contributed to generating, together with the large-scale plants, about 15 billion m^3 of biogas in 2014. It corresponds to 9 billion m^3 biomethane; moreover, the annual potential was calculated around 200–250 billion m^3 [28]. In Bangladesh, it was planned to build 100,000 small biogas systems by 2020, with an average c.a. 50 kW [36].

It is difficult for developing countries to find in the literature an exact number about the real contribution of small-scale biogas systems to the overall national renewable energy production. However, it should be noted that for the regions in which the energy access deficit is higher, domestic livestock biogas generation represents an enormous energy gain to move a step from the absolute energy poverty. For example, domestic biogas generation potential assessed in Nigeria showed an annual biogas projection of 138.7×10^6 m^3 from livestock, equivalent to 0.48 million barrels of crude oil [37].

6. Designs

6.1 Standard design systems

Biogas is a sustainable and affordable technology for rural areas where it is more convenient to adopt cheaper and simpler anaerobic systems to benefit from biogas production [38]. The household systems are low cost, simple to operate and maintain, and often constructed using local materials. The selection of the biogas systems depends on the construction, design skill, and material availability. Moreover, the design depends on the type of feedstock, climatic conditions, and geographical location. Generally, those systems do not have control instruments and heating apparatus and serve at room temperature (psychrophilic or mesophilic temperature) [5]. In tropical countries, digesters are underground to take advantage of geothermal energy; meanwhile, in mountainous regions, the systems have a reduced amount of gas to avoid discrepancies between the hot and cold season biogas production [39]. Traditionally, the generated biogas is used for cooking and lighting; however, biogas for electricity is increasing [40].

The most diffused systems in developing countries are fixed dome, floating drum, and plug flow type.

The fixed dome model is also called hydraulic digester (**Figure 1**) developed in China, where more than 45 million systems have been installed [6]; this type of system is also implemented in South Asia and Africa [31]. Typically, it consists of an underground digester and a dome-shaped roof. The digester's size depends on the amount of substrate available and the location; biodigesters are typically from 6 to 8 m^3 and operate in a semi-continuous mode. The new substrate is added once a day, while an equal amount of decanted mixed liquid is removed [5]. The digester is built from bricks, cement and reinforced by concrete. The system has one central part, the digester, dedicated to the fermentation and located at a deeper level, and above the ground level, there are two rectangular openings on each side, and they act as the inlet and outlet points for the digester. At the top of the dome-shaped roof, there is a pipe that is the biogas outlet. The digester is filled through the inlet, while the outlet also plays the hydraulic chamber's role. During the process, the biogas is produced in the digester, and it fills the upper part called the storage part (i.e., the dome). The pressure generated by the biogas presses the slurry from the digester into the inlet and outlet tanks. When the gas is released, the slurry flows back into the digester. Over the decades, this model has been improved and new

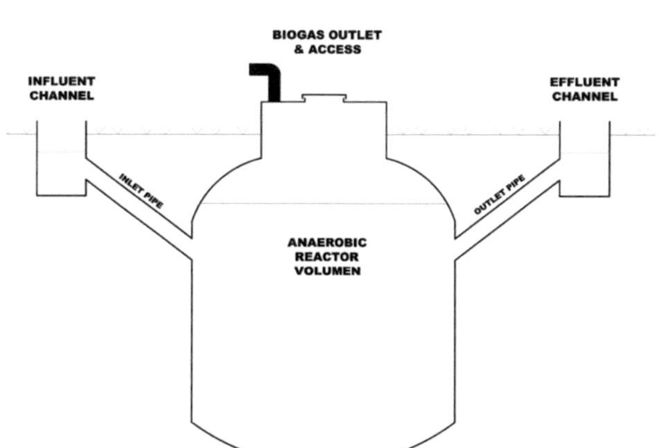

Figure 1.
Scheme of fixed dome digester model.

designs developed. In China, the digesters were modified with a hemispherical shape with a wall in the middle to increase the retention time and ensure a complete digestion process. Different fixed dome models were developed in India; first, the Janta model, a shallow system with a dome roof, has an inlet and an outlet above the dome equipped with the gas pipe. The Deenbandhu model, which is a modification of the Janta model, consists of two spheres; at the bottom, there is the fermentation unit, while at the top, there is the storage unit. In India, a low-cost model for light purposes was also designed with a vertical cylinder as a dome and with long inlet and outlet tubes [41]. In Pakistan, the French model digesters were installed; in this case, the digester is surrounded by a steel dome to prevent the loss in temperature [42]. Over the last years, alternative construction materials have been introduced to reduce labor costs and increase the system lifetime. Polymers and glass-fiber-reinforced plastics are used nowadays [43]. The fixed dome design is a reliable model with low maintenance and a long lifetime; for these reasons, it was implemented widely [31].

India developed the floating drum model (**Figure 2**); its design comprises a mobile inverted drum placed on the block digester with inlet and outlet connections through pipes located at the bottom. The digester is often partially underground. The drum acts as a reservoir; it can rise and fall along a guide pipe, depending on the produced biogas' volume. It produces biogas at constant pressure with variable volume. The weight of the drum applies the pressure required for the gas to flow through the pipeline. The digester generally is made of bricks and concrete. Meanwhile, the drum is made on metal or steel and coated with paints or bitumen to avoid corrosion, determining its lifespan. Galvanized metal and fiberglass-reinforced plastics represent a suitable alternative to standard steel [39].

The plug flow type or tubular model (**Figure 3**) was developed as portable model. This model is widespread, especially in South America [44]. It comprises a narrow and long tank (length: width equals to 5:1) inclined and partially buried in the ground, with the inlet and outlet over the ground and at the opposite side. Due to the inclination, the digestate flows toward the outlet; it is a two-phase system where acidogenesis and methanogenesis may be longitudinally separated. To keep

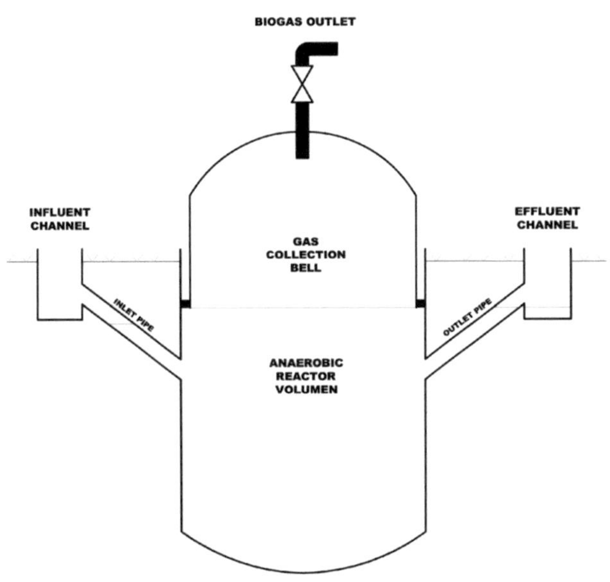

Figure 2.
Scheme of floating drum digester model.

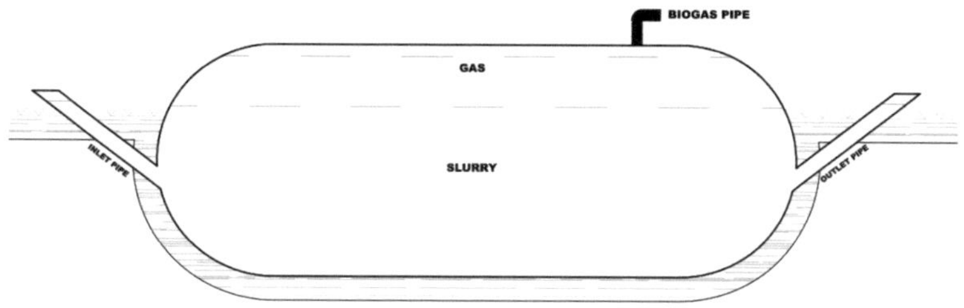

Figure 3.
Scheme of tubular digester model.

the process temperature adequate, the system needs insulation, and generally, a shed roof is placed on the top of the digester [39].

Comparing the tubular digester model with the fixed one, the fixed model can be fed with ratio manure: water 1:1, while tubular model 1:3, the former needs three times the amount of liquid [27]. Compared to the fixed dome, the plastic tubular digester has several advantages. It is a very low-cost model suitable to high altitude and low temperature, it is easy to transport and simple to install with lower investment costs, it needs less maintenance, and it is more environmentally friendly [45]. If the hard constructed models are compared from an economic point of view, for a capacity of 1–6 m^3, the cost of installation and the annual operational costs are the highest for floating model followed by fixed ones (i.e., Janta and Deenbandhu models). The floating type also has a longer payback period. With the increase of capacity, the cost of installation and the annual operational costs increase proportionally, and the payback period increases. It was shown that the Deenbandhu model (capacities from 1 to 6 m^3) is the cheapest model [46].

Regardless of the model, the household biogas systems may include auxiliary equipment to mix and handle the slurry and gas. The gas equipment may comprise pipes, valves, manometers [47].

Table 1 resumes the principal household biogas designs here described, including for each design, the advantages, the disadvantages, and the countries where it is mainly diffused.

The local conditions, biogas users' needs, waste, water, and land availability, are the criteria used to select the appropriate digester design in terms of volume and building materials [19]. Together with the different operational parameters, the design determines the biogas production and the quality of the digestate. As a decentralized energy resource, a poor design represents a particular limitation to users' adoption [50]. Moreover, sizing the digesters according to local needs and reducing the discrepancy between demand/production can avoid biogas' excessive production that often drives users to leak it into the surrounding environment purposely, and this causes a negative environmental impact [51].

6.2 Prefabricated and low-cost digesters

In recent years prefabricated systems were preferred for projects involving rural communities in developing countries. Those systems are also called "commercialized digesters" and often called "news digesters" because they involve new production materials, processes, and techniques. The main models generally used in developing countries are composite material digesters and bag digesters [9].

The bag-digester is also called balloon digester, tube digester, and it has a sealed soft plastic tubular structure. The long cylinder is generally made of polyvinyl chloride (PVC), polyethylene (PE) (**Figure 4**), or rubber. It was developed to address the construction problems with solid digesters (fixed and floating models).

Type of design	Modifications /models	Construction/Fabrication Materials	Advantages	Disadvantages	Geographical Diffusion	Ref.
Fixed dome digester	• Janta • Deenbandhu • French	• Bricks • Cement • Concrete • Polymers • Glass-fiber-reinforced plastics	• Low initial cost • long-life span (if appropriately built) • Less land required	• Requires high construction skills • Built with heavy materials • Gas leakage due to cracks	China India Nepal Uganda Tanzania	[43, 46, 48]
Floating drum digester		• Bricks and concrete for digester • Metal or Mild steel for drum • Reinforced fiber plastics • high-density polyethylene (HDPE)	• Easy construction • Visible storage volume • Gas at constant pressure	• High installation and operational costs • High payback. • Short life span (corrosion of drum) • High maintenance	India	[49]
Tubular	Pre-built and low-cost digester	• PE • PVC • HDPE • Glass fiber reinforced plastics	• Low cost • Easy transportation • Easy installation • Low maintenance	• Short life span • Requires insulation in a cold climate • Requires a high amount of water • Low gas pressure	South America Africa South Asia	[39]

Table 1.
Principal household designs used in developing countries (authors adaptation from literature sources).

Figure 4.
Example of low-cost PE- digester installation in South America. Image courtesy: Shikun Cheng.

Some Authors consider the bag digesters and the plug flow digesters different types, but actually, they are similar. In such a system, the biogas production is between 0.1 and 0.32 m^3 biogas/ m^3 digester/day, it equals the yield of traditional digesters used in India [52]. The bag-digester is more suitable in rural areas where the day temperature is above 20°C. This system has been widely applied in South and Central America [53], and at least 1 million low-cost PE plastic were installed in Vietnam with the Ministry of Agriculture and Rural Development. This system needs only two people for installation, and it can be easily transported, and for this reason, it was widely adopted for remote areas [9].

The composite material digesters are relatively new, originated in China, and well developed in East Asia countries [54]. The reinforced fiber plastic digesters represent a type of composite material digesters, and they can be manufactured through processes of resin transfer molding, sheet molding, and filament winding and they can also be built by hand (**Figure 5**). Such digesters are lightweight. Therefore, they can be easily transported and removed. They have long-term durability, good corrosion resistance to acid, high productivity, and high gas pressure (depending on the tightness). Several modified plastic digesters are also commercially available, and every model allows facile transportation. They are particularly suitable in rural

Figure 5.
Hand fabrication of composite material digester model in China. Image courtesy: Shikun Cheng.

Figure 6.
Commercial water tank (composite material digester) in Cambodia. Image courtesy: Shikun Cheng.

Figure 7.
Compact, high-rate digester for kitchen organic waste disposal. Image courtesy: Shikun Cheng.

Figure 8.
Typical portable digester for kitchen and green waste in Malaysia. Image courtesy: Shikun Cheng.

areas subject to reconstruction due to rural and land reform policy. Examples are represented by water tanks (**Figure 6**) and compact high-rate digesters (**Figure 7** and **Figure 8**) designed for kitchen and garden waste disposal [9].

7. Materials

As already mentioned in the design's description, the construction may involve different building materials. For household systems, bricks are essential material for fabricating of the digester chamber for both fixed and floating models. Generally, high-quality bricks should be used in the fabrication; however, clinker bricks are the most suitable ones because of their properties: low-cost, low moisture content, high resistivity, low thermal conductivity, appropriate thermal mass, weather resistance, fire-resistance, and tolerance to acidic pH. The concrete stones are used for building the block or the whole structure of the bricks/cement biogas digesters, they are the cheapest construction material, and they fit for the biogas purpose because of their tensile strength, durability, fire resistance, the thermal and conductive properties. The cement is also used for plastering purposes and building the concrete digester block and both the inlet and outlet. The most advantageous concrete used for the biodigesters is the Portland cement concrete (PCC), which has good density, compressive, flexural, and tensile strength. However, the use of these traditional materials brings challenges and holds disadvantages. Often the structures made with bricks, cement, and concrete, crack due to the structural stabilization and the fluctuation of temperature, usually resulting in leakages. High-quality materials and highly skilled labors are needed to minimize these problems, but those two aspects are often unavailable in developing countries. However, in recent years also alternative construction materials have been introduced like polyvinyl chloride (PVC), high-density polyethylene (HDPE), or glass fiber reinforced plastics (GRP). The PVC is used due to its low cost for building the inlet and outlet and for the digester chamber (in the case of plastic models) despite its short lifespan. Mild steel bars are usually used for the construction of the cover and the digester chamber. For the gas pipes, several different materials have been used as metal (steel or copper) and plastic (HDPE, PVC), and for the valves, generally, ball valves are used [55]. Because the biogas system's durability and cost are directly linked with construction materials, the pre-built and low-cost digesters are preferred for installations in developing countries [56]. Generally, off-site models are made with materials with specific characteristics such as glass fiber reinforced plastics (GRP), which have lower coefficient thermal conductivity, a longer operational life, and lower maintenance costs than the concrete models [54]. Several innovative design types were produced (already discussed in section 6.2), and they are commercially classified as fiber-reinforced plastic, soft plastic, and hard plastic digesters [9].

8. Influencing parameters

The process of anaerobic digestion requires the right conditions to have adequate biogas production; the most influencing parameters are temperature, organic waste composition, the moisture content, the mixing, and the hydraulic retention time (HRT) [57]. The generally suitable substrates for biogas production in rural areas are agricultural and livestock residues, organic fraction of solid domestic waste, and domestic sewage sludge (i.e., human excreta and wastewater). The biogas yield depends on the quality, amount, and supply rate (continuous or semi-continuous) of feed materials (**Table 2**). The biogas production can be directly measured by calculating the pressure of each digestor's headspace [58]. Several parameters can be used for monitoring the value of feedstocks, such as the Dry Matter (DM), the carbon-to-nitrogen ratio (C:N), Total Solids (TS), and the Volatile Solids (VS). Overall, animal manure is an ideal feedstock because of its high moisture and

Typical Feedstocks					
Manure	**Source of nutrient; high buffet capacity. Usually in co-digestion with straw.**				
Type	Organic content	DM%	VS% of DM	C:N ratio	Biogas yield [m^3/kg VS]
Pig	Carbohydrates, proteins, lipids	3–8	70–80	3–10	0.25–0.50
Cattle	Carbohydrates, proteins, lipids	5–12	80	6–20	0.20–0.30
Poultry	Carbohydrates, proteins, lipids	10–30	80	3–10	0.35–0.60
Agriculture residues	**Source of cellulose, lignin, and starch. Need pre-digestion.**				
Type	Organic content	DM%	VS% of DM	C:N ratio	Biogas yield [m^3/kg VS]
Straw	Carbohydrates, lipides	70–90	80–90	80–100	0.15–0.35
Grass		20–25	90	12–25	0.55
Organic household waste	**High variability of composition. Easily digestible. May inhibit the process for acidification.**				
Type	Organic content	DM%	VS% of DM	C:N ratio	Biogas yield [m^3/kg VS]
Fruit waste		15–20	75	15–20	0.25–50
Food residues		10	80	—	0.50–0.6

Table 2.
Common Feedstocks used in household digesters (author adaptation from literature sources).

volatile solids (VS) content and the buffering capacity, and also for its variety of microbial strains. The animal manures used in anaerobic digestion may vary according to the geographical area and local livestock practices [5, 30, 39].

The HRT always depends on temperature and substrate; however, there are no regulator instruments and no process of heating in the household systems that are generally installed in developing countries; thus, for each substrate, the optimum HRT should be found for best biogas yield because retention time affects the digestion process. The potential of cow dung, sheep, and pig manures in the plastic reactor was studied in Ethiopia, showing how at 25-28°C, a burnable gas with more than 60% of methane, was obtained from cow dung and sheep manure after 20 days of retention, while pig substrate needed more time [59]. In northern Brazil, the biogas production per kilogram of goat manure was ca. 54 L/kg in a modified floating model with a volume of 11.3 m^3 [60].

However, animal manure can make digestion slow because of its low content of carbohydrates [21], and it can generate a high concentration of ammonia, which is unfavorable for methanogens [61]. Mixing manure with other organic waste can create the optimum waste combination for the co-digestion process to improve the biomethane yield in terms of quality and quantity. Overall, the interaction within different waste streams directly determines the biogas yield [62]. In the co-digestion, the mixture of animal manure with an organic fraction rich in carbohydrates and low in ammonia has the remarkable ability to enhance biogas production. And vice versa, the agricultural residues with high VS, high fermentable constituents, and low moisture benefit from the co-digestion with animal manure or sludge due to their high content of ammonia. Compared with reactors supplied with manure alone, the volumetric methane production can increase up to 65% in reactors fed with waste and 30% VS of crop residues such as straw, sugar beet tops, and grass [63]. Co-digestion showed promising results using several mixtures of food waste and dairy manure at 35°C; a manure/food waste ratio of 52/48% produced methane yields 311 L/kg VS after 30 days of co-digestion. In comparison to raw manure, food waste contained higher VS (ca. 241 g/kg) it means higher energy content, which is desirable with regards to biogas energy production [58].

According to the different methanogenic microorganism's growth temperatures, working temperature ranges can be defined as psychrophilic (under 25°C), mesophilic (30-40°C), and thermophilic (50-60°C). Anaerobic digestion is a process that is sensitive to temperature [64]. Because simple systems as those used in rural areas in developing countries work at ambient temperature, the HRT should be selected considering local temperature conditions to give bacteria adequate time to transform feedstock into biogas. Depending on the climatic condition, the HRT varies from 10 to over 100 days [65]. At high altitude as Peruvian Andes (psychrophilic conditions), HRT from 60 to 90 days is needed [66]. In such high-altitude and cold climates, the temperature fluctuation also represents a problem for biogas production. In Andean villages, the low-cost tubular digesters were adapted by substituting the roof with a greenhouse. However, it was not always successful in maintaining a digester slurry temperature higher than the ambient temperature [64].

On the other hand, positives results were obtained from the modification of a floating drum model in Indian villages located at an altitude of 1600 to 2200 m, where the diurnal temperature fluctuates from −8 to 35°C during a year. Such fluctuation results in the reduction of gas production during winter by 23–37%. An improvement of the insulation kept proper operating temperature. That was achieved by enfolding the system inside a greenhouse or using hollow bricks for the construction or placing straw insulation around the digester, or adding hot water in the input feedstock material. These modifications allowed a continuous biogas production around 1.6 to 2.6 m^3/day during the whole year [67]. Solar-biogas hybrid systems where a solar collector provided the heating have been proposed for maintaining the right temperature for anaerobic bacteria to produce biogas [68].

In tropical regions with mesophilic conditions, the HRT may range from 20–60 days [19]. In Bangladesh, the rural dome-type digesters showed a retention time of about 40–50 days from a single feedstock such as cows' manure [29]. In Nigeria, the total biogas produced from poultry and cassava wastes was 1.5 m^3 after 42 days in a prototype polyethylene system of 1 m^3 at the ambient temperature of 33.6°C [69].

It is important to retain that while the temperature will affect the biogas, the feedstock security (or availability) influences the operation of the system [70]. For fueling a household stove twice per day in a family of five persons, it is required manure from one pig, five cows, or 130 chickens to have approximately 1.5 m^3 of biogas [6]. Gathering sufficient water and manure are among the limiting factors; in many parts of Sub-Saharan Africa, although the households possess adequate livestock, the grazing nature (nomadic, semi-nomadic, or free) may impede to gather manure to feed the biogas digesters [71]. A digester volume of 1.3 m^3/capita requires approximately 0.05 m^3/day of water for each cow and 0.01 m^3/day for each pig supplying manure to the digester. Such an amount of water can hardly be provided in areas of low water availability. In sub-Saharan countries, the water needed for digestion can be provided using recycled waters (gray water), such as domestic water, rainwater harvesting and aquaculture [72].

All rural small-scale and household digesters models require daily operation and maintenance. Everyday operations include the feeding, the handling of digestate, and the control of biogas outflow. Both brick and plastic tubular digesters are supplied with organic waste diluted with water in different proportions. The most challenging maintenance for the users comprises removing sludge from the digester, blocking possible cracks in the fixed digesters, and repairing damages in plastic systems [19]. Because installed digesters' functionality depends on continuous management and supervision of operation and maintenance, specific programs are often put in place to develop ownership and participation in using the biogas systems [73]. Sensitivity analyses demonstrated that small-sized digesters are more environmentally sustainable, if biogas leakage and release are avoided [51].

9. The relevance of small-scale biogas systems to regional development of rural areas in developing countries

The literature study discloses how small-scale biogas systems benefit the local family, village, and surrounding communities in rural areas in developing countries. Anaerobic digestion, even at the small-scale, represents an efficient waste treatment, and it offers a source of clean energy (biogas) suitable for cooking, heating, electricity generation, and a digestate with a high fertilizer value. It is a widespread opinion that anaerobic digestion implemented in poor rural areas may help in achieving several Sustainable Development Goals (SDGs), positive health impacts and sanitization, preservation of soil and water [74], reduction of greenhouses gas (GHG) emissions, gender empowerment and education [75], and accessible and affordable source of clean energy [76].

The use of biodigesters to treat human sludge and animal manure significantly improves the hygiene situation of rural areas that lack adequate infrastructure to collect and treat wastewater, unmanaged human and animal waste. The use of biodigesters can reduce infectious diseases such as diarrhea, cholera, and tuberculosis. Biodigesters also reduce the environmental impact (ecological, health, esthetic) of the spreading of waste in rural areas and reduce sewage danger percolating into the groundwater sources pumped for drinking water. Moreover, it contributes to the reduction of GHG emissions. It was calculated that processing the liquid and solid manure through anaerobic digestion reduces the potential impact from 4.4 kg carbon dioxide (CO_2) equivalents to 3.2 kg CO_2 equivalents if compared with traditional manure management [77].

Biodigesters represent a great alternative to the inefficient use of traditional biomass such as fuelwood, agricultural residues, and dried dung. Rural areas worldwide suffer from the loss of forest lands due to the illegal collection of firewood. The installation of biodigesters and the use of biogas can provide a substitute for firewood and save forests. Also, fuel oil and kerosene are widely used in rural areas for cooking and lighting purposes, especially in developing countries. Biogas is an excellent replacement for these fossil fuels and can save people hundreds of dollars every year. Besides that, countries with large amounts of rural areas are usually poor and oil-importing countries. The use of biogas can save those countries millions of dollars every year.

The use of biogas as a clean source of energy for cooking also includes important health benefits. It reduces exposure to indoor smoke and soot, reduces respiratory and eye diseases, reduces fatalities caused by carbon monoxide poisoning and offers a significant reduction of the RSPM in indoor environments.

Biogas use has many positive social outcomes on education and gender equality, and it generates employment opportunities for rural communities. The lack of enough lighting in rural areas in developing countries prevents students of all ages from having enough light to study or even be involved in any educational activities in the evenings. Biogas in gas lamps provide enough fuel for lighting and provide more study hours in the dark [78]. Moreover, in such poor areas, women are in charge of securing water and energy [67, 75, 79]. Having a biodigester at home will save women tens of hours of collecting firewood. This time can be used by women for other activities such as education and socializing. Also, burning biogas does not generate any particulate matter or soot that pollutes the houses, saving women cleaning time [21, 78]. Moreover, an increase in employment in rural areas was recognized as the positive impact of small-scale biogas installations. These news opportunities mainly involved women and professionals in education, environment, agriculture, and technical professions related to the building and maintenance of the systems.

The use of biodigesters reduces the use of chemical fertilizers. Along with the biogas, biodigester produces organic fertilizer rich in nutrients, such as nitrogen, potassium, and phosphorus. This organic fertilizer can replace commercial fertilizers and save farmers in rural areas thousands of dollars every year. Also, this liquid fertilizer can keep the use of water for irrigation. Thus, biodigesters maximize the valuable fertilizing properties of the recycled waste for agriculture; this benefit will lead and promote the local family's economic advancement.

10. Biogas serves to reduce energy poverty in developing countries

In some countries, rural people do not even have access to fossil oil and kerosene because of their price or shortage; those people are forced to meet their energy needs using traditional and inefficient resources. As described, such practices represent significant health, environmental, economic, and social issues for those communities. Within the context of sustainable development, nowadays, it is imperative to offer these disfavored regions access to clean, affordable, and renewable energy. Assisting people to transform the animal manure, crop residues, domestic waste into a more efficient energy carrier, such as biogas, provide clean and reliable energy, and conserve the local and global environment [21]. It is evident how biogas' decentralized production gives several opportunities for accelerating the transition to sustainable development and the circular economy with positive economic effects at the local-level livelihood [80]. Biogas is an energy source useful for people to meet their energy needs without using fossil fuel [8].

In Northern Brazil, a biogas volume of 1 m^3 from manure was equal to 0.75 L of gasoline [60]. Small-scale biodigesters produce around 2–4 m^3/day biogas, sufficient to meet the cooking lighting needs of a family [62]. The biogas potential in Colombia showed that 80% of propane, which is used the traditional fuel, could be replaced by biogas; results showed that a low-cost tubular digester in polyethylene with a total volume of 9.5 m^3 and feed with cattle produces enough biogas to supply cooking of five hours/day for five people [81]. In India, positive achievements were obtained using different design models simultaneously; it was possible to produce approximately 40.5 m^3 biogas/day and supply the community of 48 households that had cooking needs of 0.85 m^3/day each [82]. In Bangladesh, about eight head of cattle per household were needed to cover the need for cooking gas, electricity, and drinking water [83]. In Nepal, 0.33 m^3 of biogas fulfills the energy needs per capita per day [84]. In Israel, post-nomadic Bedouins families adopted a system of 7.5 m^3 fueled with goat manure and straw that provided biogas for cooking and for powering a little refrigerator [85]. In Bali approximately 30 m^3 biogas/month using cow manure can supply the energy need of a 5–6 people family size [86].

Small-size biogas technology embodies the opportunity to address the energy access issue for low-income developing countries [87]. Biogas digesters may reduce energy poverty [35, 88], and they provide clean energy for cooking and lighting for rural areas where energy infrastructures are missing [39].

11. Challenges of biogas systems in rural areas for communities in developing countries

Despite all of the benefits biodigesters have for rural communities, some biogas systems in rural areas do not meet the expectations due to technology, maintenance, and technical support. All those aspects induce a discontinuity of digester operation as documented for China, in the Guizhou Province, 62.03% of household biogas were continuously operating while 36.72% were discontinued [89]. In some other cases, the challenges represent the reasons for technology's abandonment [90]. This section summarizes the challenges biogas systems are facing in rural areas.

In cold rural areas, biogas system owners lack the right technology to maintain the thermal conditions for a high rate of biogas production [57]. The people in these areas face this challenge, especially in winters where energy need is higher than in other seasons. As described above, the household biogas digesters are made of bricks or concrete and built just under the ground surface where the digesters' temperature is very close to the ambient temperatures. Thus, without appropriate heating or hybrid technologies, the household biodigesters' efficiency remains low and unstable under these conditions. Design solutions have been developed to maintain the right temperature for biogas production, such as insulating the digesters or combining with other heating technology (i.e., solar water heaters). However, these solutions may cause a burden for people in rural areas.

The lack of technical knowledge and building capacity in rural areas is another critical factor that leads to low biogas production rates. People in rural areas lack access to formal education, awareness of environmental issues, agricultural techniques, and appropriate knowledge on how to run the biodigesters. In some countries, farmers get governmental financial supports to construct biogas systems. In many cases, this governmental support is not accompanied by technical support and safety measures to adequately manage the biodigesters [21, 26, 78, 91]. Also, the lack of knowledge about the ratio between the size of the biodigester and the volume of organic waste can lead to low biogas production rates and digestate pollution near the biodigester. That may cause odor emissions, eutrophication of surface water, and pollution of groundwater. As described below, only a rational design of the small-scale system, along with a proper build, continuous cleaning, and maintenance, affects the productivity and the environmental footprint of the system [51].

In general, rural areas are located in remote zones where it is difficult to reach and run educational programs and maintenance. Also, the lack of governmental follow-up and capacity building programs leads to poor maintenance and operation of the biogas plants.

The inadequate use of liquid fertilizer may attract flies and mosquitoes to the biodigester and cause a challenge for the biodigester users. Also, this may create adverse publicity of biogas plants among people.

Low or discontinuous biogas production due to improper operation of the biodigester, technical barriers, lack of feedstock (animal manure or food waste), and low level of awareness may lead to an inadequate supply of biogas. Thus, people in rural areas are discouraged from using the biodigesters on a daily or seasonal basis. It may lead to low adoption rates in rural areas and force people to switch to more reliable fuel sources.

12. Conclusion

The chapter presents the effective implementation of small biogas digesters in rural areas in developing countries. Small Biogas digesters represent a tool to achieve rural areas' sustainable development, giving access to clean and affordable renewable energy. The use of biodigesters in poor rural areas serves as an environmentally friendly way to reduce fossil fuels and traditional biomass and reduce indoor and outdoor air pollution. Also, the use of biogas can significantly reduce organic waste in poor rural areas. Design, construction materials, feedstock operational modes vary accordingly with the geographical location of biogas installation. The systems installed in rural areas are simple and mainly for domestic uses. The biogas yield can be controlled and increased by controlling the retention time and modulating feedstock composition in a co-digestion process using manure and other organic waste. Despite the potential and the wide range of benefits that rural areas can acquire from the small-biogas digesters, several potential problems limit the diffusion of small-scale anaerobic digesters in rural areas in developing countries. They include the lack of construction and maintenance skills, awareness of users, and the inadequacy of design to meet the actual biogas (energy) need. For biogas systems to succeed and be used in rural areas worldwide, governments should strengthen current policies and develop new policies and regulations to motivate people in rural areas to install biodigesters. These policies should focus on the comprehensive sustainability of the biogas systems. The policies should include incentives and procedures for constructing the biogas digesters and comprise tools to support the systems' management.

Acknowledgements

The authors wish to thank Dr. Shikun Cheng for the kind permission to use prefabricated biodigesters photos.

Conflict of interest

The authors declare no conflict of interest.

Author details

Martina Pilloni[1*] and Tareq Abu Hamed[2,3]

1 University of Cagliari, Cagliari, Italy

2 Arava Institute for Environmental Studies, Kibbutz Ketura, Israel

3 Dead Sea and Arava Science Center, Israel

*Address all correspondence to: martina.pilloni@gmail.com

References

[1] Salunkhe, D.B.; Rai, R.K.; Borkar, R.P. Biogas Technology. *Int. J. Eng. Sci. Technol.* **2012**, *4*, 4934-4940.

[2] Nevzorova, T.; Kutcherov, V. Barriers to the wider implementation of biogas as a source of energy: A state-of-the-art review. *Energy Strateg. Rev.* **2019**, *26*, 100414, doi:10.1016/j.esr.2019.100414.

[3] Chynoweth, D.P.; Owens, J.M.; Legrand, R. Renewable methane from anaerobic digestion of biomass. *Renew. Energy* **2001**, *22*, 1-8, doi:10.1016/S0960-1481(00)00019-7.

[4] Edelmann, W.; Schleiss, K.; Joss, A. Ecological, energetic and economic comparison of anaerobic digestion with different competing technologies to treat biogenic wastes. *Water Sci. Technol.* **2000**, *41*, 263-273, doi:10.2166/wst.2000.0080.

[5] Teodorita Al Seadi, Domiik Rutz, Heinz Prassl, Michael Kottner, Tobias Finsterwalder, Silke Volk, R.J. *Biogas handbook*; University of Southern Denmark, 2008; ISBN 9788799296200.

[6] Bond, T.; Templeton, M.R. History and future of domestic biogas plants in the developing world. *Energy Sustain. Dev.* **2011**, *15*, 347-354, doi:10.1016/j.esd.2011.09.003.

[7] Laramee, J.; Davis, J. Economic and environmental impacts of domestic bio-digesters: Evidence from Arusha, Tanzania. *Energy Sustain. Dev.* **2013**, *17*, 296-304, doi:10.1016/j.esd.2013.02.001.

[8] Amini, S.; Ordoney, P.; Khuzestan, I. Investigation of biogas as renewable energy source. *Int. J. Agric. Crop Sci.* **2013**, 1453-1457

[9] Cheng, S.; Li, Z.; Mang, H.P.; Huba, E.M.; Gao, R.; Wang, X. Development and application of prefabricated biogas digesters in developing countries. *Renew. Sustain. Energy Rev.* **2014**, *34*, 387-400, doi:10.1016/j.rser.2014.03.035.

[10] Bilsborrow, R.E. Population growth, internal migration, and environmental degradation in rural areas of developing countries. *Eur. J. Popul.* **1992**, *8*, 125-148, doi:10.1007/BF01797549.

[11] Mulinda, C.; Hu, Q.; Pan, K. Dissemination and Problems of African Biogas Technology. *Energy Power Eng.* **2013**, *05*, 506-512, doi:10.4236/epe.2013.58055.

[12] Bruun, S.; Jensen, L.S.; Khanh Vu, V.T.; Sommer, S. Small-scale household biogas digesters: An option for global warming mitigation or a potential climate bomb? *Renew. Sustain. Energy Rev.* **2014**, *33*, 736-741, doi:10.1016/j.rser.2014.02.033.

[13] Kanagawa, M.; Nakata, T. Analysis of the energy access improvement and its socio-economic impacts in rural areas of developing countries. *Ecol. Econ.* **2007**, *62*, 319-329, doi:10.1016/j.ecolecon.2006.06.005.

[14] Reddy, A.K.N. Lessons from the Pura community biogas project. *Energy Sustain. Dev.* **2004**, *8*, 68-73, doi:10.1016/S0973-0826(08)60468-8.

[15] Kamp, L.M.; Bermúdez Forn, E. Ethiopia's emerging domestic biogas sector: Current status, bottlenecks and drivers. *Renew. Sustain. Energy Rev.* **2016**, *60*, 475-488, doi:10.1016/j.rser.2016.01.068.

[16] Tock, L.; Schummer, J. Sustainable waste-to-value biogas plants for developing countries. *Waste Manag.* **2017**, *64*, 1-2, doi:10.1016/j.wasman.2017.05.014.

[17] Mshandete, A.M.; Parawira, W. Biogas technology research in selected

sub-Saharan African countries – A review. *African J. Biotechnol.* **2009**, *8*, 116-125, doi:10.5897/AJB08.821.

[18] Cho, J.C.; Cho, H.B.; Kim, S.J. Heavy contamination of a subsurface aquifer and a stream by livestock wastewater in a stock farming area, Wonju, Korea. *Environ. Pollut.* **2000**, *109*, 137-146, doi:10.1016/S0269-7491(99)00230-4.

[19] Ferrer-Martí, L.; Ferrer, I.; Sánchez, E.; Garfí, M. A multi-criteria decision support tool for the assessment of household biogas digester programmes in rural areas. A case study in Peru. *Renew. Sustain. Energy Rev.* **2018**, *95*, 74-83, doi:10.1016/j.rser.2018.06.064.

[20] World Energy Outlook 2020 – Analysis - IEA Available online: https://www.iea.org/reports/world-energy-outlook-2020 (accessed on Feb 14, 2021).

[21] Surendra, K.C.; Takara, D.; Hashimoto, A.G.; Khanal, S.K. Biogas as a sustainable energy source for developing countries: Opportunities and challenges. *Renew. Sustain. Energy Rev.* **2014**, *31*, 846-859, doi:10.1016/j.rser.2013.12.015.

[22] Ortiz, W.; Terrapon-Pfaff, J.; Dienst, C. Understanding the diffusion of domestic biogas technologies. Systematic conceptualisation of existing evidence from developing and emerging countries. *Renew. Sustain. Energy Rev.* **2017**, *74*, 1287-1299, doi:10.1016/j.rser.2016.11.090.

[23] He, P.J. Anaerobic digestion: An intriguing long history in China. *Waste Manag.* **2010**, *30*, 549-550, doi:10.1016/j.wasman.2010.01.002.

[24] Giwa, A.S.; Ali, N.; Ahmad, I.; Asif, M.; Guo, R.B.; Li, F.L.; Lu, M. Prospects of China's biogas: Fundamentals, challenges and considerations. *Energy Reports* **2020**, *6*, 2973-2987, doi:10.1016/j.egyr.2020.10.027.

[25] Lichtman, R. Toward the diffusion of rural energy technologies: Some lessons from the Indian biogas program. *World Dev.* **1987**, *15*, 347-374, doi:10.1016/0305-750X(87)90018-0.

[26] Mittal, S.; Ahlgren, E.O.; Shukla, P.R. Barriers to biogas dissemination in India: A review. *Energy Policy* **2018**, *112*, 361-370, doi:10.1016/j.enpol.2017.10.027.

[27] Martí-Herrero, J.; Chipana, M.; Cuevas, C.; Paco, G.; Serrano, V.; Zymla, B.; Heising, K.; Sologuren, J.; Gamarra, A. Low cost tubular digesters as appropriate technology for widespread application: Results and lessons learned from Bolivia. *Renew. Energy* **2014**, *71*, 156-165, doi:10.1016/j.renene.2014.05.036.

[28] Scarlat, N.; Dallemand, J.F.; Fahl, F. Biogas: Developments and perspectives in Europe. *Renew. Energy* **2018**, *129*, 457-472, doi:10.1016/j.renene.2018.03.006.

[29] Khan, E.U.; Martin, A.R. Review of biogas digester technology in rural Bangladesh. *Renew. Sustain. Energy Rev.* **2016**, *62*, 247-259, doi:10.1016/j.rser.2016.04.044.

[30] Vasco-Correa, J.; Khanal, S.; Manandhar, A.; Shah, A. Anaerobic digestion for bioenergy production: Global status, environmental and techno-economic implications, and government policies. *Bioresour. Technol.* **2018**, *247*, 1015-1026, doi:10.1016/j.biortech.2017.09.004.

[31] Ghimire, P.C. SNV supported domestic biogas programmes in Asia and Africa. *Renew. Energy* **2013**, *49*, 90-94, doi:10.1016/j.renene.2012.01.058.

[32] Association, W.B. Global Bioenergy Statistics 2019. *World Bioenergy Assoc.* **2019**, 58.

[33] Wang, X.; Yan, R.; Zhao, Y.; Cheng, S.; Han, Y.; Yang, S.; Cai, D.; Mang, H.P.; Li, Z. Biogas standard system in China. *Renew. Energy* **2020**, *157*, 1265-1273, doi:10.1016/j.renene.2020.05.064.

[34] Mutungwazi, A.; Mukumba, P.; Makaka, G. Biogas digester types installed in South Africa: A review. *Renew. Sustain. Energy Rev.* **2018**, *81*, 172-180, doi:10.1016/j.rser.2017.07.051.

[35] Amigun, B.; Blottnitz, H. von Investigation of scale economies for African biogas installations. *Energy Convers. Manag.* **2007**, *48*, 3090-3094, doi:10.1016/j.enconman.2007.05.009.

[36] Bertsch, N.; Marro, P. Making Renewable Energy a Success in Bangladesh : Getting the Business Model Right Making Renewable Energy a Success in Bangladesh : **2015**.

[37] Adeoti, O.; Adegboyega, T.D.; Ayelegun, T.A. An assessment of Nigeria biogas potential from agricultural wastes. *Energy Sources* **2001**, *23*, 63-68, doi:10.1080/00908310151092173.

[38] Woldeyohannes, A.D.; Woldemichael, D.E.; Baheta, A.T. Sustainable renewable energy resources utilization in rural areas. *Renew. Sustain. Energy Rev.* **2016**, *66*, 1-9, doi:10.1016/j.rser.2016.07.013.

[39] Rajendran, K.; Aslanzadeh, S.; Taherzadeh, M.J. *Household biogas digesters-A review*; 2012; Vol. 5; ISBN 4633435485.

[40] Ciotola, R.J.; Lansing, S.; Martin, J.F. Emergy analysis of biogas production and electricity generation from small-scale agricultural digesters. *Ecol. Eng.* **2011**, *37*, 1681-1691, doi:10.1016/j.ecoleng.2011.06.031.

[41] Jash, T.; Basu, S. Development of a mini-biogas digester for lighting in India. *Energy* **1999**, *24*, 409-411, doi:10.1016/S0360-5442(98)00102-9.

[42] Nazir, M. Biogas plants construction technology for rural areas. *Bioresour. Technol.* **1991**, *35*, 283-289, doi:10.1016/0960-8524(91)90126-5.

[43] Deng, L.; Liu, Y.; Zheng, D.; Wang, L.; Pu, X.; Song, L.; Wang, Z.; Lei, Y.; Chen, Z.; Long, Y. Application and development of biogas technology for the treatment of waste in China. *Renew. Sustain. Energy Rev.* **2017**, *70*, 845-851, doi:10.1016/j.rser.2016.11.265.

[44] Garfí, M.; Martí-Herrero, J.; Garwood, A.; Ferrer, I. Household anaerobic digesters for biogas production in Latin America: A review. *Renew. Sustain. Energy Rev.* **2016**, *60*, 599-614, doi:10.1016/j.rser.2016.01.071.

[45] Garfí, M.; Cadena, E.; Pérez, I.; Ferrer, I. Technical, economic and environmental assessment of household biogas digesters for rural communities. *Renew. Energy* **2014**, *62*, 313-318, doi:10.1016/j.renene.2013.07.017.

[46] Singh, K.J.; Sooch, S.S. Comparative study of economics of different models of family size biogas plants for state of Punjab, India. *Energy Convers. Manag.* **2004**, *45*, 1329-1341, doi:10.1016/j.enconman.2003.09.018.

[47] Vögeli, Y.; Riu, C.; Gallardo, A.; Diener, S.; Zurbrügg, C. *Anaerobic Digestion of Biowaste in Developing Countries*; 2014; ISBN 9783906484587.

[48] Walekhwa, P.N.; Mugisha, J.; Drake, L. Biogas energy from family-sized digesters in Uganda: Critical factors and policy implications. *Energy Policy* **2009**, *37*, 2754-2762, doi:10.1016/j.enpol.2009.03.018.

[49] Kandpal, T.C.; Joshi, B.; Sinha, C.S. Economics of family sized biogas plants in India. *Energy Convers. Manag.* **1991**, *32*, doi:10.1016/0196-8904(91)90150-H.

[50] Yaqoot, M.; Diwan, P.; Kandpal, T.C. Review of barriers to the dissemination

of decentralized renewable energy systems. *Renew. Sustain. Energy Rev.* **2016**, *58*, 477-490, doi:10.1016/j.rser.2015.12.224.

[51] Ioannou-Ttofa, L.; Foteinis, S.; Seifelnasr Moustafa, A.; Abdelsalam, E.; Samer, M.; Fatta-Kassinos, D. Life cycle assessment of household biogas production in Egypt: Influence of digester volume, biogas leakages, and digestate valorization as biofertilizer. *J. Clean. Prod.* **2021**, *286*, 125468, doi:10.1016/j.jclepro.2020.125468.

[52] Lansing, S.; Víquez, J.; Martínez, H.; Botero, R.; Martin, J. Quantifying electricity generation and waste transformations in a low-cost, plug-flow anaerobic digestion system. *Ecol. Eng.* **2008**, *34*, 332-348, doi:10.1016/j.ecoleng.2008.09.002.

[53] Garfí, M.; Ferrer-Martí, L.; Perez, I.; Flotats, X.; Ferrer, I. Codigestion of cow and guinea pig manure in low-cost tubular digesters at high altitude. *Ecol. Eng.* **2011**, *37*, 2066-2070, doi:10.1016/j.ecoleng.2011.08.018.

[54] Nguyen, V.C.N. Small-scale anaerobic digesters in Vietnam – development and challenges. *J. Vietnamese Environ.* **2011**, *1*, 12-18, doi:10.13141/jve.vol1.no1.pp12-18.

[55] Obileke, K.; Onyeaka, H.; Nwokolo, N. Materials for the design and construction of household biogas digesters for biogas production: A review. *Int. J. Energy Res.* **2020**, 1-18, doi:10.1002/er.6120.

[56] Garfí, M.; Castro, L.; Montero, N.; Escalante, H.; Ferrer, I. Evaluating environmental benefits of low-cost biogas digesters in small-scale farms in Colombia: A life cycle assessment. *Bioresour. Technol.* **2019**, *274*, 541-548, doi:10.1016/j.biortech.2018.12.007.

[57] Sommer, S.G. Optimising simple biogas digesters for use in cold regions of developing countries. *Biotechnology* **2007**, 115-120.

[58] El-Mashad, H.M.; Zhang, R. Biogas production from co-digestion of dairy manure and food waste. *Bioresour. Technol.* **2010**, *101*, 4021-4028, doi:10.1016/j.biortech.2010.01.027.

[59] Gemechu, F.K. Evaluating the Potential of Domestic Animal Manure for Biogas Production in Ethiopia. *J. Energy* **2020**, *2020*, 1-4, doi:10.1155/2020/8815484.

[60] Borges Neto, M.R.; Carvalho, P.C.M.; Carioca, J.O.B.; Canafístula, F.J.F. Biogas/photovoltaic hybrid power system for decentralized energy supply of rural areas. *Energy Policy* **2010**, *38*, 4497-4506, doi:10.1016/j.enpol.2010.04.004.

[61] Fujino, J.; Morita, A.; Matsuoka, Y.; Sawayama, S. Vision for utilization of livestock residue as bioenergy resource in Japan. *Biomass and Bioenergy* **2005**, *29*, 367-374, doi:10.1016/j.biombioe.2004.06.017.

[62] Bharathiraja, B.; Sudharsana, T.; Jayamuthunagai, J.; Praveenkumar, R.; Chozhavendhan, S.; Iyyappan, J. Biogas production – A review on composition, fuel properties, feed stock and principles of anaerobic digestion. *Renew. Sustain. Energy Rev.* **2018**, *90*, 570-582, doi:10.1016/j.rser.2018.03.093.

[63] Lehtomäki, A.; Huttunen, S.; Rintala, J.A. Laboratory investigations on co-digestion of energy crops and crop residues with cow manure for methane production: Effect of crop to manure ratio. *Resour. Conserv. Recycl.* **2007**, *51*, 591-609, doi:10.1016/j.resconrec.2006.11.004.

[64] Alvarez, R.; Lidén, G. The effect of temperature variation on biomethanation at high altitude. *Bioresour. Technol.* **2008**, *99*, 7278-7284, doi:10.1016/j.biortech.2007.12.055.

[65] Gunnerson, C.; Stuckey, D. *Integrated Resource Recovery-Anaerobic Digestion*; 1986; ISBN 0821307525.

[66] Ferrer, I.; Garfí, M.; Uggetti, E.; Ferrer-Martí, L.; Calderon, A.; Velo, E. Biogas production in low-cost household digesters at the Peruvian Andes. *Biomass and Bioenergy* **2011**, *35*, 1668-1674, doi:10.1016/j.biombioe.2010.12.036.

[67] Lohan, S.K.; Dixit, J.; Kumar, R.; Pandey, Y.; Khan, J.; Ishaq, M.; Modasir, S.; Kumar, D. Biogas: A boon for sustainable energy development in India's cold climate. *Renew. Sustain. Energy Rev.* **2015**, *43*, 95-101, doi:10.1016/j.rser.2014.11.028.

[68] Wang, D.; Duan, Q.; Li, Y.; Tian, X.; Rahman, S. Simulation of a solar-biogas hybrid energy system for heating, fuel supply, and power generation. *Int. J. Energy Res.* **2017**, *41*, 1914-1931, doi:10.1002/er.3754.

[69] Ezekoye, V.A.; Ezekoye, B.A.; Offor, P.O. Effect of Retention Time on Biogas Production from Poultry Droppings and Cassava Peels. *Nig J. Biotech* **2011**, *22*, 53-59.

[70] Naik, L.; Gebreegziabher, Z.; Tumwesige, V.; Balana, B.B.; Mwirigi, J.; Austin, G. Factors determining the stability and productivity of small scale anaerobic digesters. *Biomass and Bioenergy* **2014**, *70*, 51-57, doi:10.1016/j.biombioe.2014.01.055.

[71] Mwirigi, J.; Balana, B.B.; Mugisha, J.; Walekhwa, P.; Melamu, R.; Nakami, S.; Makenzi, P. Socio-economic hurdles to widespread adoption of small-scale biogas digesters in Sub-Saharan Africa: A review. *Biomass and Bioenergy* **2014**, *70*, 17-25, doi:10.1016/j.biombioe.2014.02.018.

[72] Bansal, V.; Tumwesige, V.; Smith, J.U. Water for small-scale biogas digesters in sub-Saharan Africa. *GCB Bioenergy* **2017**, *9*, 339-357, doi:10.1111/gcbb.12339.

[73] Tigabu, A.D.; Berkhout, F.; van Beukering, P. Technology innovation systems and technology diffusion: Adoption of bio-digestion in an emerging innovation system in Rwanda. *Technol. Forecast. Soc. Change* **2015**, *90*, 318-330, doi:10.1016/j.techfore.2013.10.011.

[74] Breitenmoser, L.; Gross, T.; Huesch, R.; Rau, J.; Dhar, H.; Kumar, S.; Hugi, C.; Wintgens, T. Anaerobic digestion of biowastes in India: Opportunities, challenges and research needs. *J. Environ. Manage.* **2019**, *236*, 396-412, doi:10.1016/j.jenvman.2018.12.014.

[75] Yasar, A.; Nazir, S.; Tabinda, A.B.; Nazar, M.; Rasheed, R.; Afzaal, M. Socio-economic, health and agriculture benefits of rural household biogas plants in energy scarce developing countries: A case study from Pakistan. *Renew. Energy* **2017**, *108*, 19-25, doi:10.1016/j.renene.2017.02.044.

[76] SEforALL SEforALL Analysis of SDG7 Progress - 2020. **2020**.

[77] Vu, T.K. V; Vu, D.Q.; Jensen, L.S.; Sommer, S.G.; Bruun, S. Life cycle assessment of biogas production in small-scale household digesters in Vietnam. *Asian-Australasian J. Anim. Sci.* **2015**, *28*, 716-729, doi:10.5713/ajas.14.0683.

[78] Gautam, R.; Baral, S.; Herat, S. Biogas as a sustainable energy source in Nepal: Present status and future challenges. *Renew. Sustain. Energy Rev.* **2009**, *13*, 248-252, doi:10.1016/j.rser.2007.07.006.

[79] Katuwal, H.; Bohara, A.K. Biogas: A promising renewable technology and its impact on rural households in Nepal. *Renew. Sustain. Energy Rev.*

2009, *13*, 2668-2674, doi:10.1016/j.rser.2009.05.002.

[80] Lyytimäki, J. Renewable energy in the news: Environmental, economic, policy and technology discussion of biogas. *Sustain. Prod. Consum.* **2018**, *15*, 65-73, doi:10.1016/j.spc.2018.04.004.

[81] Castro, L.; Escalante, H.; Jaimes-Estévez, J.; Díaz, L.J.; Vecino, K.; Rojas, G.; Mantilla, L. Low cost digester monitoring under realistic conditions: Rural use of biogas and digestate quality. *Bioresour. Technol.* **2017**, *239*, 311-317, doi:10.1016/j.biortech.2017.05.035.

[82] Singh, A.K.; Kaushal, R.K. Design of Small Scale Anaerobic Digester for Application in Indian Village : A Review. *J. Eng. Appl. Sci.* **2016**, 11-16.

[83] Khan, E.U.; Mainali, B.; Martin, A.; Silveira, S. Techno-economic analysis of small scale biogas based polygeneration systems: Bangladesh case study. *Sustain. Energy Technol. Assessments* **2014**, *7*, 68-78, doi:10.1016/j.seta.2014.03.004.

[84] Centre for Energy Studies Institute of Engineering Pulchok, Lalitpur, N. *Biogas Support Programme (BSP) and Netherlands Development Organiza- tion (SNV/N). Advanced courses in bio-gas technology. Lalitpur, Nepal*; 2001;

[85] Pilloni, M.; Hamed, T.A.; Joyce, S. Assessing the success and failure of biogas units in Israel: Social niches, practices, and transitions among Bedouin villages. *Energy Res. Soc. Sci.* **2020**, *61*, 101328, doi:10.1016/j.erss.2019.101328.

[86] IRENA; Heezen, P.A.M.; Gunnarsdóttir, S.; Gooijer, L.; Mahesh, S.; Ghiandelli, M. Measuring small-scale biogas capacity and production. *Chem. Eng. Trans.* **2016**, *31*, 37-42, doi:10.3303/CET1331007.

[87] Somanathan, E.; Bluffstone, R. Biogas: Clean Energy Access with Low-Cost Mitigation of Climate Change. *Environ. Resour. Econ.* **2015**, *62*, 265-277, doi:10.1007/s10640-015-9961-6.

[88] Antoine Halff, Benjamin K. Sovacool, and J.R. *Energy Poverty: Global Challenges and Local Solutions*; Halff, A., Sovacool, B.K., Rozhon, J., Eds.; Oxford University Press, 2014; ISBN 9780199682362.

[89] Wang, X.; Lu, X.; Yang, G.; Feng, Y.; Ren, G.; Han, X. Development process and probable future transformations of rural biogas in China. *Renew. Sustain. Energy Rev.* **2016**, *55*, 703-712, doi:10.1016/j.rser.2015.09.097.

[90] Lwiza, F.; Mugisha, J.; Walekhwa, P.N.; Smith, J.; Balana, B. Dis-adoption of Household Biogas technologies in Central Uganda. *Energy Sustain. Dev.* **2017**, *37*, 124-132, doi:10.1016/j.esd.2017.01.006.

[91] Jiang, X.; Sommer, S.G.; Christensen, K. V. A review of the biogas industry in China. *Energy Policy* **2011**, *39*, 6073-6081, doi:10.1016/j.enpol.2011.07.007.

Chapter 5

Techno Economic Studies on the Effective Utilization of Non-Uniform Biowaste Generation for Biogas Production

Godwin Glivin, Mariappan Vairavan,
Premalatha Manickam and Joseph Sekhar Santhappan

Abstract

Environmental effects from traditional energy sources and government regulations, necessitate the use of alternative energies like biogas for many uses including drying and refrigeration. Biowaste produced in educational institutions will not be uniform over the year. The non-uniform supply of biowastes, the absence of studies on bio digestion of likelihood biomass, the unreliability of energy from such conversion and the profitability of its usage in most applications are some of the factors to be considered while implementing this technology. In this regard, theoretical and experimental evaluations were carried out to accurately forecast biogas generation capabilities in educational campuses for obtaining biofuels with quantity and efficiency. It is observed that biogas generation with 52 to 58% methane content can be possible during an academic year. The quality of biogas shows that it is appropriate for almost any application. A broader analysis on different types of biogas digesters was conducted for their suitability in academic institutions. The economic benefits are analyzed for incorporating three biogas digesters namely KVIC, Fiber Reinforced Plastic (FRP) type and JANATA. There are some encouraging results to confirm the economic feasibility of biogas plants including positive net present value. Biogas generation with digesters of capacities varying between 25 and 450 cubic meter shows payback periods varies from 3.18 to 7.59 years, which confirms that it is profitable to use digesters in this range of capacities.

Keywords: biogas, biodigester types, economic analysis, payback period, non-uniform loading rates

1. Introduction

1.1 Renewable energy: current scenario

The environmental factors and depletion of conventional energy sources create a huge demand for technologies to substitute conventional fuels. Renewable Energy Sources (RES) such as solar, wind, tidal and biomass are available abundantly and they can be harvested without environmental degradation. The International Energy Outlook (IEO) states that the global primary energy demand will increase to

48% between 2012 and 2040 [1]. The share of non-renewable energy (liquid fuels, coal, natural gas and nuclear) will decrease from 91% in 1990 to 84% in 2040. However, renewable energy sources will continue to grow and catering from 9% of the world's energy demand to 16%. The share of primary energy sources in the world's energy generation also points a decrease in the non-renewable energy's share in electricity generation from 78–71% in 2040.

The growth of installed capacity of renewable energy sources in India shows that the country had gone up from 7.8% in 2008 to 15.9% in 2016 with the generation mix of wind power (57%), solar power (18%), biomass (15%), small hydro (9%) and waste to energy (1%). Waste to energy is one of the new classifications among the energy mixes in the country. Among the various renewable energy conversion technologies, biochemical conversion is one of the best techniques to convert biowaste to useful form of energy (biogas). This low-cost technology can convert any organic wastes to biogas which can be further used as a fuel for cooking, lighting, power generation, etc. Anaerobic Digestion (AD) is one of the RES conversion processes which is capable of handling 90% of moisture content [2]. The end product of the AD is biogas which is comprised mainly with CH_4 and CO_2. CH_4 is the combustible gas with an energy content of 50 ± 5 MJ/kg which can be utilized for heating, power generation and other applications related with gaseous fuel [3].

The AD process involves four steps (hydrolysis, acidogenesis, acetogenesis and methanogenesis) which is effected by methanogens such as hydrogenotrophic and acidogenic [4]. The organic content consists of various particulate as well as water insoluble polymers, hence the polymers are not accessible for the microorganisms directly [5, 6]. During the first step i.e., hydrolysis the insoluble polymers break down to soluble oligomer and monomer. This is caused by the strains of hydrolytic bacteria which releases hydrolytic enzymes [7]. Carbohydrates, lipids, and proteins are converted to sugars, long-chain fatty acids, and amino acids. In the next step i.e., acidogenesis the soluble molecules are converted to $C0_2$ and H_2 along with acetic acid, propionic acid, ethanol, and alcohols. Other acids which are produced apart from acetic acid, propionic acid, ethanol are due to *Actinomyces, Peptostreptococcus anaerobius, Clostridium* and *Lactobacillus* respectively [8]. With the support of proton reducing agent the long volatile fatty acids as well as alcohols will oxidize to acetic acid and H_2 during acetogenesis (third step) [9]. During the last stage (methanogenesis) methanogens are generated namely *hydrogenotrophic* and *acetoclastic* [10, 11]. This is caused by the reduction of CO_2 to H_2 as well as scrubbing of sliced acetic acid which is formed in the third stage. The biochemical conversion process involved in the AD is shown in **Figure 1**.

1.2 Biogas production and utilization

The data obtained from the year-wise installed capacity in MW of bio-power energy sources for power generation in India reveals that the installed capacity of bio-power energy sources has been on the increase every year and the same can be utilized for about 70% of the rural basic energy needs in India [12]. Bio-power produced by thermochemical (biomass gasification) and biochemical (biogas) conversion techniques contributes significantly to India's rural energy supply. According to a 2012 World Bank report, waste is classified as organic, paper, plastic, bottles, metals, among others. For most solid waste preparation purposes, these six categories are normally appropriate. Studies in the field of biowaste utilization in Europe showed high initial cost for the implementation; however, such cost could be reduced by intensive research on process integration and intensification. The ministry of MNRE, India has set a target of 10 GW of bio-power capacity by 2022 [13]. A huge potential is observed for employing anaerobic digestion as waste

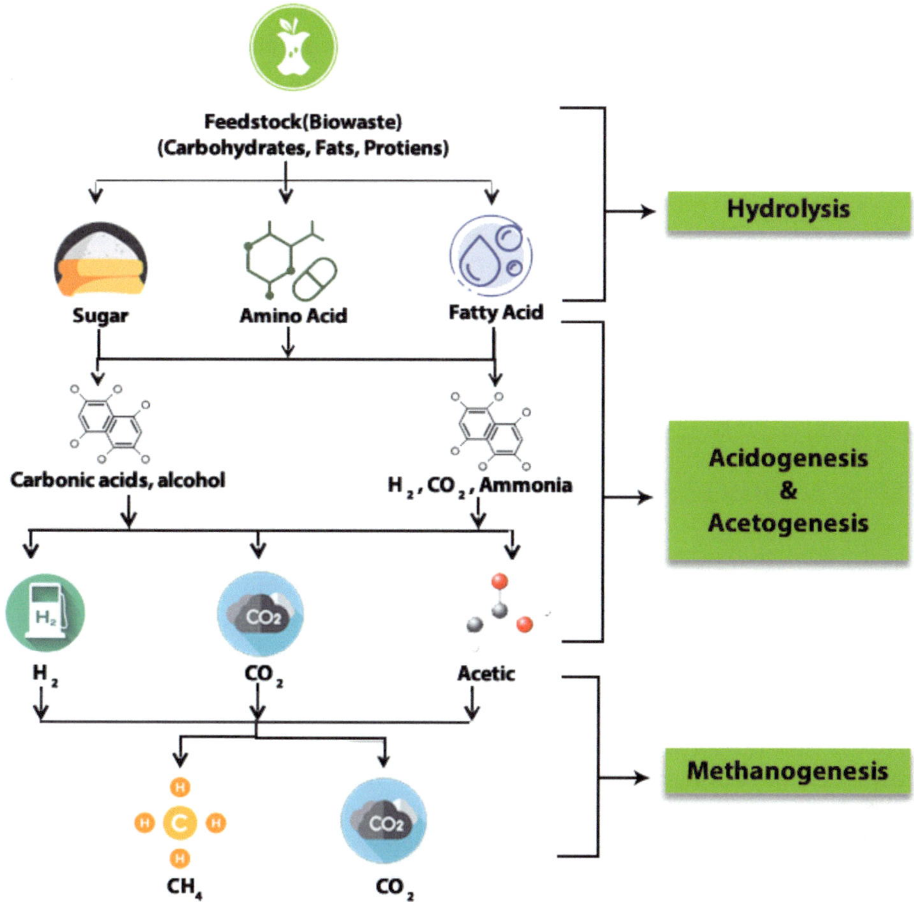

Figure 1.
Anaerobic digestion process.

management method and energy production technology in India and the rest of the world [14].

Realizing the potential of biogas as future energy source, many studies were conducted on biogas generation, utilization, and applications. The canteen and mess wastes which are rich in organic content could be used effectively for waste utilization and energy generation. The series of experiments conducted by varying HRT and OLR showed that with at Hydraulic Retention Time (HRT) of 20 days and 100 kg TS m^3d^{-1}, the methane content of 50% with 0.981m^3 kg^{-1} VS could be achieved [15]. A test conducted with mesophilic tubular digester for generation of biogas showed that fruits and vegetable wastes were used as feedstock. Variations in HRT and feed concentration were used to assess the digester's efficiency. With a feed concentration of 6% TS and a 20-day HRT, the digester's efficiency was found to be the highest [16]. An experiment was conducted with pig manure in Anaerobic Batch Reactor (ABR) for hydrogen generation in two stages for pH values 5.0, 5.5 and 6.0. The OLR was taken as 96.4, 48.2 and 32.1 kg VS $m^{-3}d^{-1}$ whereas HRT was maintained as 12, 24 and 36 h. It was noted that at 12 h HRT and 96.2 kg VS $m^{-3}d^{-1}$ OLR, the hydrogen concentration was at the maximum [17].

An analysis was carried out to check the stability and performance of anaerobic digestion with varying HRT and OLR. The analysis showed a decrease in methane yield with the increase in OLR as well as a decrease in HRT for low OLR

(0.1 g $VS^{-1} d^{-1}$). At high HRT (25 days), the methane yield was maximum [18]. Co-digestion of food waste and fruit-vegetable waste was performed in single-phase and two-phase digesters. By varying the OLR, authors concluded that single-phase digester could produce more methane than two-phase for low OLR [19]. According to reports, co-digesting food waste with cattle manure will boost biogas production and methane yield [20]. The performance of biodigesters under overload conditions was evaluated based on two case studies. To study the interrelation between biomass population dynamics and digester stability, Anaerobic Digestion Model 1 (ADM1) was utilized. The study showed that the digester did not function in high OLR conditions [21]. The techno- economic study of a combined bioprocess, based on solid state fermentation for fermented hydrogen generation from food waste was conducted. The outcome shows that five years Pay Back Period (PBP), 26.75 percent Return on Investment (ROI) and 24.07 percent and Internal Rate of Return (IRR) respectively could be possible [22].

1.3 Scope and aims of the work

Many studies reported the production and utilization of biogas for various applications. In most of them, technical and economic viability of biogas plants for the utilization of biogas in various applications was studied for a stable organic loading in biodigesters. Despite the high potential for biogas use in educational facilities, only a few studies have been conducted to determine the techno-economic feasibility of using biogas technology in this field [23–25]. This is mainly due to the variation of student and staff population throughout a year, and the non-uniform generation of organic waste. Furthermore, in order to improve the accuracy of the forecast, the quality and quantity of biogas produced from various biowastes available in this area must be investigated. Hence, this current research focuses on predicting technological and economic influences, as well as their effect on the deployment of biogas plants in a few educational institutions in India's southern region. The following objectives have been established to scientifically research the feasibility of using biowastes available in educational institutions in the selected area, as well as to determine the effect of non-uniform loading on digester's efficiency and economic viability.

- Identify and characterize the biowaste available in educational institutions.
- Find the impacts of non-uniform loading of biowastes on the biogas generation in biodigesters using mathematical and experimental methods.
- Predict the economic factors for the implementation of biogas digesters in a few educational institutions.

2. Methodology

2.1 Grouping of biowastes and selection of biogas plants

Anaerobic digestion based waste management technology has an enormous significance in India because of the vital role of waste disposal methods as well as its role as a renewable energy source for cooking, lighting, electricity generation, and so on [26]. The anaerobic digestion process utilizes a variety of biowastes from various sources including municipal solid waste, households, institutions, and industry. The generation of biogas from anaerobic digestion of biowaste in educational institutions is projected to play a significant role in ensuring rural and urban prosperity [27]. As a result, institutions in and around the southern part of India were chosen for this research, where biogas will substitute 35 percent to 40 percent of the traditional fuel used for cooking. The institutions in this region were categorized based on the student population, and the potential of biowastes and their availability throughout a year were studied. The strategy followed to select the biowaste and the digestion systems has been shown in **Figure 2**.

2.2 Categorization of institutions

More accurate research is possible in educational institutions because the large number of students living in the campus offers numerous opportunities for biogas production. Based on the population of students and staff, the institutes situated in southern part of India (the region selected for this study) were categorized as A, B, C, D and E as mentioned in **Table 1**. The population details were collected based on the published data of the respective institution.

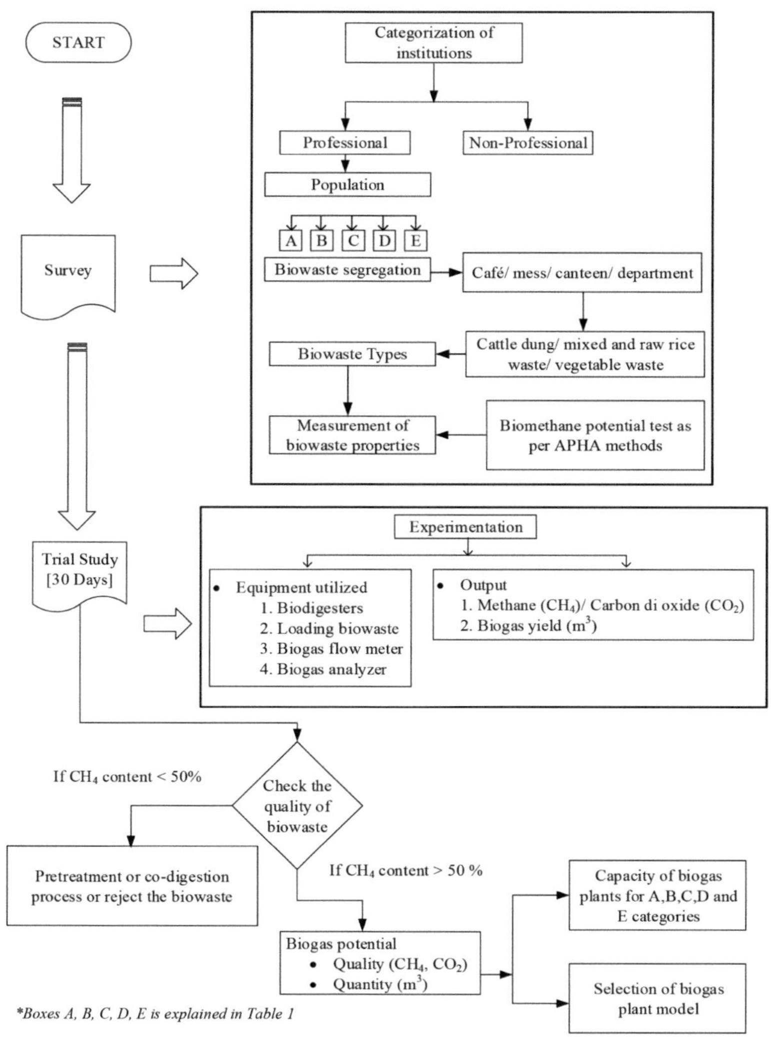

Figure 2.
Flow chart for the procedure involved in the grouping of biowastes and the selection of biogas plants.

Categories	Range of population	Institutions in numbers	Population mean
A	1000–2500	200	1728
B	2501–5000	180	3448
C	5001–9000	95	6399
D	9001–20,000	75	11,500
E	20,001–40,000	20	29,231

Table 1.
The various categories of institutions according to population range.

2.3 Selection of biowastes for this study

A survey was conducted with the required questionnaire to select the biowaste samples. Biowaste details such as amount, consistency, and varieties were discovered through the survey. The type of institution, academic schedule, population of students and staff living on and off campus, biowaste generation sources, conventional cooking fuel, and other relevant factors dominated questionnaire's development. Personal information of people was also included. The data reliability was verified with relevant authorities.

2.3.1 Potential of biowaste sources

Sewage sludge (SS), food waste (FW), leaves, cotton waste, paper waste, and other biogas energy sources have been reported. **Table 2** shows the estimated data of a sample.

On a regular basis for different academic schedules, a survey on food waste supply in a group 'A' institution was performed. This research looked at the most traditional food menu trends used by different institutions. Food wastes produced before and after cooking were also taken into account. **Table 3** shows the specification of category 'A' institution.

Table 4 shows the common biowastes and the percentage of biowaste generated in a category 'A' institution. The samples were collected in the hostels before dumping. Separate buckets were kept for collecting the different food wastes. The students and staff members were instructed to dump the leftover food accordingly. It was observed that the availability of some wastes like fruit waste, meat waste and fish waste was low but their quantity in total waste had been checked at least twice a

Particulars	Category 'A' Institution
Geographical area of the institution (acres)	27–35
Total population	1000–2500
Literacy of population (%)	100
Density of livestock population	0–15
Waste disposal technology	Landfill, open heating
Biowaste suitable for anaerobic digestion (kg/day)	100–700
Quantity of dung production (kg/day)	0–70
Quantity of Convention fuel (LPG) used for cooking (kg/day)	12–15

Table 2.
The data grouped for a category 'A' institution.

Particulars	Sample Institution
Geographical area of the institution (acres)	34
Population range	200–2400
Literacy of population (%)	100
Density of livestock population	15
Waste disposal technology	Landfill
Biowaste suitable for anaerobic digestion (kg/day)	100–590
Quantity of dung production (kg/day)	45–70
Quantity of Convention fuel (LPG) used for cooking (kg/day)	12

Table 3.
The data grouped for a sample category 'A' institution.

Sl. no.	Biowastes	kg of biowastes
1.	Cooked rice	44
2.	Cooked vegetables	3.7
3.	Tea	2.8
4.	Coffee	2.2
5.	Salad	3.7
6.	Oil	11.2
7.	Fruit wastes	16.9
8.	Mixed rice wastes	490

Table 4.
Sample data for biowastes generated in a category 'A' institution on 100th day.

month to find any major deviation. The observation showed that the variation was not significant. Hence such wastes were added along with mixed rice waste.

Among the numerous biowastes generated in the study area, Rice Waste (RW), Mixed Rice Waste (MRW), and Vegetable Waste (VW) were some of the potential biowastes available. Therefore, they were selected for the anaerobic digestion. Meat, fish, potato, and rice wastes, left out after consuming were used in MRW. **Table 5** shows the grouped-biowastes used as feedstock for biogas generation. Other biowastes, apart from VW and RW, were mixed with MRW due to insufficient availability.

2.4 Measurement of biowaste properties

The important parameters which control biogas generation are pH, VS and TS, therefore these properties were experimentally measured as per the standard procedure discussed below [28].

Sl. No.	Biowastes	kg
1.	Rice waste	5–50
2.	Mixed rice wastes	70–490
3.	Vegetable waste	5–50

Table 5.
Biowastes grouping for category 'a' institution.

2.4.1 Total solids

The following technique was used to assess the feed's TS according to APHA guidelines [28]. 50 g of each biomass was placed in pre-weighed porcelain vessels and heated at 60°C for 24 hours and then at 103°C for 3 hours in a hot air oven. The weight of the dry samples, as well as the container, was determined in a weighing balance with a precision of 0.001 g. A sample's TS percentage was determined as follows:

$$TS = \left(\frac{W_d}{W_w}\right) \cdot 100 \tag{1}$$

The dry and wet sample weights are W_d and W_w, respectively.

2.4.2 Volatile solids

The standard formula for determining the VS of feed materials was used. The oven-dried samples were dried at 550°C ± 50°C and ignited fully inside the muffle furnace. The desiccator's cooled samples were measured, and VS was determined using the Eq. (2).

$$VS = \left[\frac{(W_d - W_a)}{W_a}\right].100 \tag{2}$$

where W_d is the dry sample weight, and W_a is the dry ash weight.

2.4.3 pH

The pH of biowastes Cow Dung (CD), RW, MRW, and VW was measured at least once in a day using a pH electrode with 0.05 percent accuracy. The samples were taken from the slurry until where it was fed to the digesters. A pH electrode dipped in the inoculum was used to test pH of digesters on daily basis. **Table 6** shows chemical properties of the four types of biowastes used in this study. Eqs. (1) and (2) were used to measure the values of TS and VS. The validity of experiments was verified after the findings were compared to literature.

2.5 Biogas plants commonly used in India

In India, more than seven models of biogas plants are available and they are being used in various parts of the country according to the requirement of a particular area [35]. This study examines the feasibility of applying appropriate model in educational institutions from Khadi and Village Industries Commission (KVIC),

Sl. No.	FEED	pH %		TS %		VS %	
		Current study	Reference values	Current study	Reference values	Current study	Reference values
1	CD	6.50	6.30 [29]	15.98	17 [29]	64.99	89 [29]
2	MRW	4.91	4–7 [30]	20.25	14.4 [31]	90.15	89.5 [31]
3	RW	6.61	4–7 [30]	30.28	14.4 [31]	90.11	89.5 [31]
4	VW	6.35	7.1 [32]	10.55	9.3 [33]	90.45	78–93 [34]

Table 6.
Characterization of feedstock.

JANATA, and Fiber-Glass Reinforced Polyester (FRP) [36]. These three models were selected based on the ease in construction as well as operation compared with other models. The selection of biogas plant model varies for all institutions based on the nature and activities of the students. For selected category of institutions these three models were considered.

2.5.1 Biogas Plant: Khadi and Village Industries Commission

This type of biogas plants consists of a floating drum made of steel, fiber glass reinforced polyester or high-density polyethylene. Its underground digester tank is made of bricks and cement as shown in **Figure 3**. The floating drum which moves up and down according to the biogas generation serves as the gas holder. The major disadvantage of these models is high maintenance due to corrosion of drum which leads to regular coatings. The rainwater should be prevented from entering the tank as it corrodes the steel. The advantage is seen when the same model floating drum is made of fiber glass reinforced polyester or high-density polyethylene, it can work efficiently without affecting the digestion process but it makes the biogas plant more expensive. The life of the plant is found to be 15 years [37].

2.5.2 Biogas plant: JANATA

The fixed dome instead of the floating drum, as seen in **Figure 4**, distinguishes this from KVIC model. Initial cost of dome is lower than that of KVIC model since it is constructed by bricks, blocks, and cement. The major disadvantage of this model is making a gas tight dome because in such models, leaks are observed in the cracks formed in the dome due to poor construction. Thus, this type of biogas plants required skilled supervisors and labourers for construction. This kind of small-scale biogas plant has a lower cost, making it a good choice for institutions in categories A, B, and C. A long life of 20 years or more can be expected due to non-corrosive parts used in construction [37]. Compared with other two models, this model has the largest life span.

2.5.3 Biogas plant: fiber-glass reinforced polyester

The FRP model biogas plants as shown in **Figure 5** are most used in household applications in both rural and semi-urban parts of India. FRP is used in the

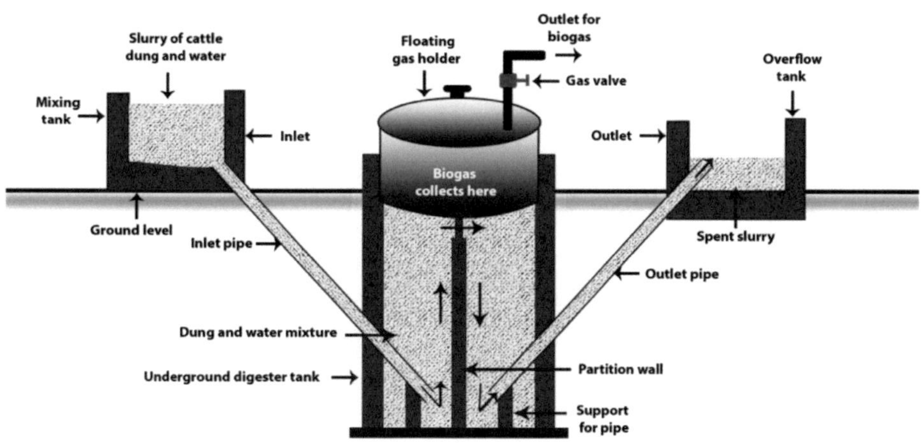

Figure 3.
Biogas plant with floating-drum and cylindrical digester (KVIC model).

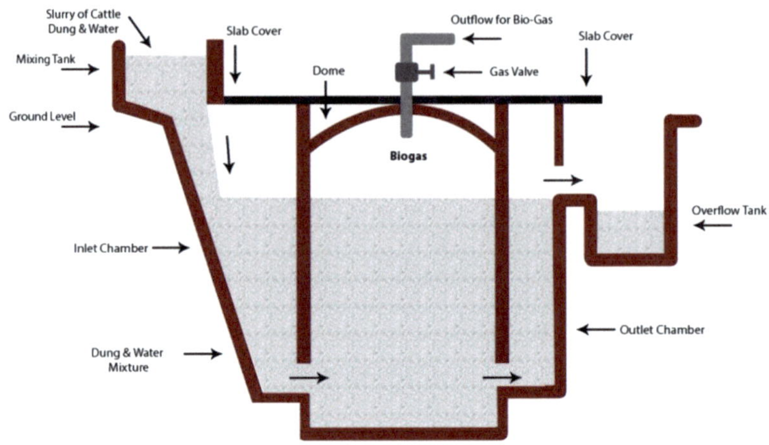

Figure 4.
Brick-reinforced fixed-dome biogas plant (JANATA model).

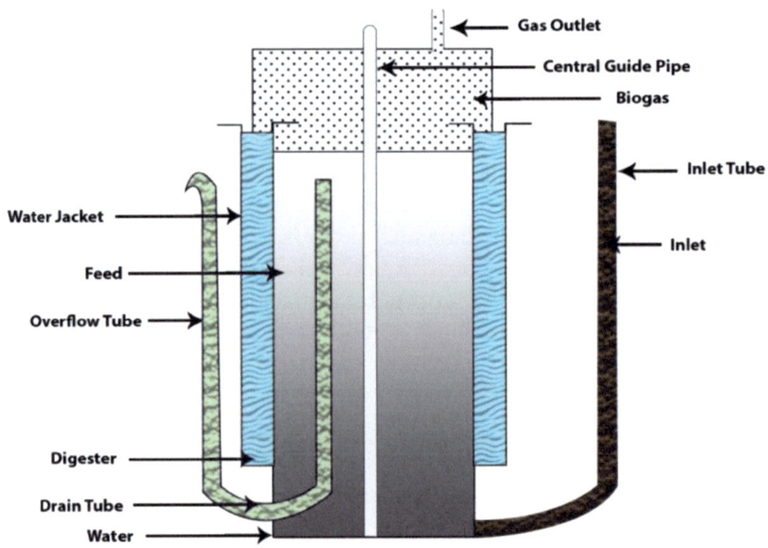

Figure 5.
Biogas plant with floating drum made by fiberglass reinforced polyester.

construction of digester tank, floating drum, and water jacket. PVC pipes are used for inlet and outlet pipes, and the central guide pipe is made of GS. Unlike other models, these biogas plants are placed above earth due to smaller in size. The maximum size of this type of biogas plants is limited to 1 to 12 m^3. The FRP model biogas plants are portable and can be easily maintained. The investment cost is less, and such models are more attractive for small scale applications. The space occupied by this model is one of the disadvantages compared with other two models. An average of 10 year life span has been reported for this model [23].

2.6 Mathematical modeling

Educational institution is a place where the generation of biowaste is high during academic schedule whereas low in non-academic schedules. This non-uniformity in biowaste availability affects the loading rate which results in reduced methanogens activity. Hence, by understanding the performance of digesters with available biowastes throughout a year, the minimum and maximum production of biogas in various academic schedules can be predicted. Further, it can be used to design the capacity of a biogas plant toward efficiently manage the variations in daily yield. As part of a theoretical simulation, a study was conducted to predict biodigesters' efficiency and their effect on non-uniform loading. The equations that state the mathematical representation of biochemical reactions are used for the analysis in Anaerobic Digestion Model 1 (ADM1). Therefore, ADM1 toolbox was adopted to represent the complete metabolic network of an anaerobic digestion [11]. This toolbox aids in determining the system's operational conditions as well as its behaviour. Moreover, it could help in the design of biogas plants of large scale.

The various steps used for the simulation are depicted in **Figure 6**. The simulation process starts with the selection of biowastes for anaerobic digestion. The properties such as pH, TS, VS, and moisture content (MC) of biowaste were studied through APHA procedures and taken as input parameters [38, 39]. The temperature levels, digester tank scale, and simulation phase were chosen from the respective inbuilt parameter control menus. Then the simulation was carried out in steps of a day, and the quality and quantity of the biowaste were measured. If the measured quality of methane was less than 50% the biowaste was rejected and a new one was selected for the simulation.

2.7 Experimental setup

Figure 7 shows a schematic diagram of the experimental system included in the analysis. It holds a digester tank which is surrounded by a water jacket. The floating drum, known as gas holder, is fixed in such a way that it can move up and down based on the generation of biogas. The water jacket holds the floating drum and prevents the

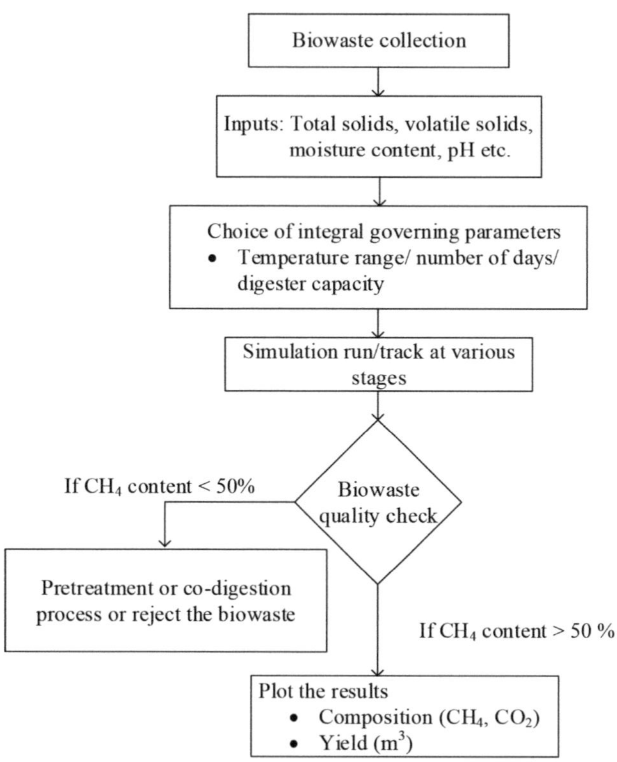

Figure 6.
Flow chart of the simulation procedure.

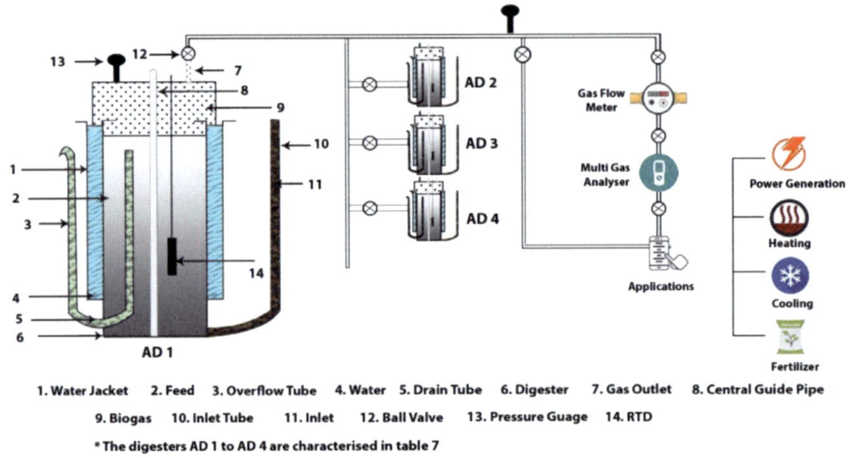

Figure 7.
Schematic diagram of the experimental setup.

leakage of biogas and odor of inoculum. A stainless-steel central guide is mounted in the centre of the digester tank to ensure smooth flow of the floating drum. To load biowaste and drain digestate, inlet and outlet pipes are provided appropriately. Drainpipes are also provided to clean the digester tank and water jacket. Suitable arrangement is made in the floating drum to transfer the biogas for any application.

To calculate the quantity and consistency of the biogas, a thermal gas flow metre (mass flow measurements of liquids) with a 0.5 percent Full Scale (F.S) accuracy and a multi gas analyzer (NUCON) with 0.3 percent accuracy are attached in the gas line. A pH electrode and temperature sensors are dipped inside the inoculum. The manifold connects all the digesters with the instrumentation panel.

2.7.1 Experimental procedure

Initially Cow Dung (CD) was filled in all the four digesters for the generation of methanogens with an HRT of 55 days. After confirming the complete digestion of CD, the required quantity of biowastes collected from the educational institution of category A was loaded for 30 days with the same quantity per day. The quality and quantity of methane generated per day was measured using the multi gas analyzer and thermal gas flow meter. The pH and temperature of the feedstock during digestion process were also measured at regular intervals and their averages were calculated. During this trial study, the temperature was observed between 29–34°C. **Figure 8** depicts a photographic image of the digesters used in the experimental setup as mentioned in **Table 7**.

After the trial study the same digesters were used for the pilot study for 365 days. However, the loading was varied according to the non-uniformity in the availability of biowastes. Since the total quantity of biowastes generated inside the campuses cannot be digested completely with the small digesters, only 10% of each type of waste was taken every day and the same was used for loading the digester. Thus, the impact of non-uniform generation of biogas was incorporated in the pilot study. The results were used in the prediction of quality and quantity of biogas generated for the proposed systems.

2.8 Economic study

The economic feasibility of a biogas plant for non-uniform loading is also important to confirm the selection of any type. As a result, the economic study was

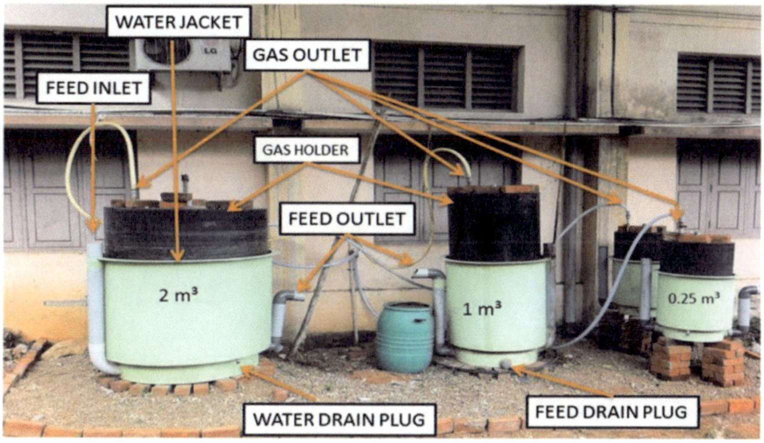

Figure 8.
Experimental setup of different capacity biogas digesters.

Digester	Capacity (m^3)	Feed	Slurry ratio (Feed: water)	Temperature/State	Period (days)
AD1	2	MRW	1:1	32 ± 5 °C/Mesophilic	365
AD2	1	CD	1:1	32 ± 5 °C/Mesophilic	365
AD3	0.25	RW	1:1	32 ± 5 °C/Mesophilic	365
AD4	0.25	VW	1:1	32 ± 5 °C/Mesophilic	365

Table 7.
Summary of the experimental design.

done using Capital Cost (CC), Annual Operating Cost (AOC), Payback Period (PBP), Net Present Value (NPV), and Life Cycle Cost (LCC). For this study, standard equations from previous studies have been chosen [37, 40]. Based on the pilot study performed in category 'A' institution, the biogas produced per person per day was determined and found vary from 0.014 to 0.019 m^3. A mean value of 0.015 m^3 per person was taken into consideration. Methane content was found as 53%. The capacity of the biogas plant for each category was calculated using the mean value. The quality and quantity of biogas generated over the course of a year were also determined using primary data.

The biogas plant's volume (size) for an institution is determined by the availability of biowaste and the biogas yield from it. Using data from a pilot study conducted in category "A" institution, the supply of biowaste in the other categories of institutions over the span of a year was calculated and plotted in **Figure 9**. It is observed that the capacity of the biogas plant for each category varies between 25 m^3 and 450 $m^{3.}$ The calculations were carryout based on the average values taken from the population range as mentioned in **Table 1**. Hence, different types of biogas plants are required for each institution based on certain parameters such as geographical location, climatic condition, transportation and so on. Hence, the specifics of the different biogas plants available in India were investigated.

2.8.1 Selection of biogas plants in an economic analysis

The three types of biogas plants namely KVIC, JANATA and FRP were considered in this economic analysis. These models were selected based on the geographic

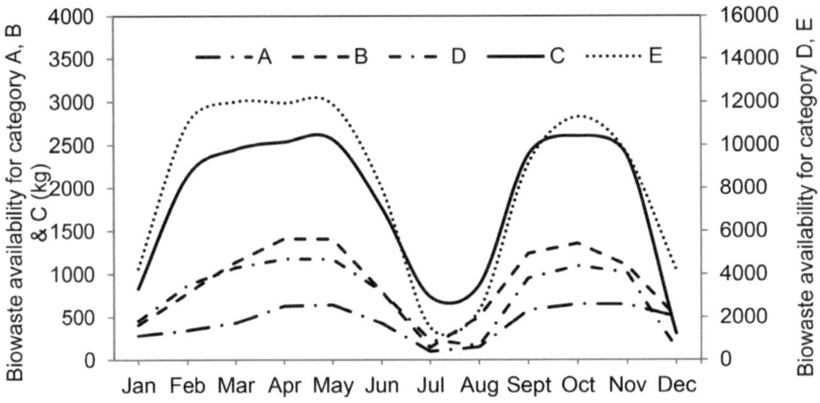

Figure 9.
Estimation of the biowaste availability for all categories.

location and the capacity of waste in an institution. Because of the simplicity of design and construction, KVIC models are the best choice for higher capacity biogas plants. The KVIC model plants suffer from a disadvantage in hilly areas because of the rusting in floating drum according to various climatic changes. JANATA model biogas plants, on the other hand, which are entirely made of bricks, resist rusting and are thus strongly recommended. Due to portability feature, FRP models are highly suggested for less capacity requirement. The initial investment is one of the major concerns for these types of biogas plants. Due to such concerns, the various economic factors are studied and discussed below.

2.8.1.1 Capital cost

The cost of the digester, construction costs, and government subsidies are all included in the CC of the Biogas Plant (BGP). Eq. (3) is used to calculate the capital expenditure.

$$\text{Capital Cost} = \text{Cost of the biogas plant} + \text{Installation cost of biogas plant} \quad (3)$$

2.8.1.2 Running cost

The operating and repair costs as well as the annual depreciation value, contributes to the plant's running expense. The cost of maintenance is estimated to be 2% of the plant's capital cost. (Jatinder & Sarbjit, 2004). For KVIC, JANATA, and FRP models, the life span was assumed as15, 20, and 10 years, respectively. The measurements are dependent on a handling fee of Rs 0.40 per kg for biowaste, which covers shipping and labour costs.

$$\begin{aligned}\text{Running Cost} = {} & \text{Cost of the biowaste used} \\ & + \text{cost of maintenance and operation of biogas plant} \\ & + \text{cost of manpower/labour} + \text{transportation charge} \\ & + \text{depreciation value} \quad (4)\end{aligned}$$

2.8.1.3 Payback period

The economics of a biogas plant includes the calculation of the payback period to substitute the LPG cooking stoves with biogas-based cooking stoves. It has been calculated as

$$Payback\ period = \frac{Cost\ of\ Installation}{Annual\ Profit} \tag{5}$$

Where, Annual profit is the difference between the annual income and the annual operational cost of the BGP.

2.8.1.4 Net present value

The present value of a system's spending and operating costs over its lifespan is known as the net present value (NPV). NPV is one of the main economic factors for comparing the energy conversion systems. The difference between the present value of the benefits and the costs resulting from an investment is the net present value of the investment. It is calculated by,

$$NPV = \left[S.\left(\frac{(1+i)^n - 1}{i(1+i)^n}\right)\right] - CC \tag{6}$$

Where, 'S' - benefits at the end of the period, CC - initial capital investment, i - annual interest rate (12%).

The below are the approval conditions for an investment project as determined by the NPV method:

a. accept the system if NPV > 0

b. reject the system if NPV < 0

2.8.1.5 Life cycle cost

Another significant economic metric is the system's LCC, which accounts for all expenses involved with the system over its lifetime by considering the worth of money. The Life Cycle Cost Analysis (LCCA), which considers the initial costs, operation costs, repair costs, replacement costs, and salvage prices, is a valuable method for determining whether the selected biogas plants could be installed in educational institutions. A life cycle of 15, 20 and 10 years were assumed in calculating the Present Worth Cost (PWC) of KVIC, JANATA and FRP biogas plants [41].

$$\text{LCC} = \text{Initial costs} + \text{POC} + \text{PMC} + \text{PRE} + \text{PSV} \tag{7}$$

where, POC – present worth cost of the operating cost. PMC– present worth cost of the maintenance cost. PRE– present worth cost of the replacement cost. PSV– present worth cost of the salvage value.

Parameter (INR)	Relation
Annual operation cost (AOC)	Energy source cost + running (operation as well as maintenance) cost + depreciation value
Income From Gas (IFG)	Cost of LPG per kg * Equivalent of 1 LPG
Income From Slurry (IFS)	0.3 * Annual dung requirement
Total Income (TI)	IFG + IFS
Annual profit	TI - AOC

Table 8.
The relations used to calculate selected economic parameters.

Parameters	Value	Reference
Annual O&M cost (INR/year)	2% of CC	[37]
Annual interest rate (%)	12	[42]
NPV (evaluation period in years)	KVIC (15) JANATA (20) FRP (10)	[42–44]
LCC (life span in years)	KVIC (15) JANATA (20) FRP (10)	[42–44]

Table 9.
Economic parameters for the analysis.

Tables 8 and **9** lists the several parameters that are incorporated in the economic analysis.

3. Results and discussion

3.1 Pilot study: influence of non-uniform loading rate

The non-uniform generation of biowaste in an educational institution for 365 days was studied to check the performance in terms of methane content and biogas yield. To understand the different academic schedules the study period has been divided into four phases as mentioned in **Table 10**.

According to academic schedules, the biowaste generation per day during maximum population was found as 70 kg, 280 kg, 120 kg and 80 kg for CD, MRW, RW and VW, and during minimum population it was 70 kg, 120 kg, 60 kg and 20 kg respectively. 10% of each biowaste was taken for the loading throughout a year as shown in **Figure 10**.

The biogas yield was observed for all the biowastes during different phases according to the loading pattern. To study the deviation of this biogas yield from uniform loading, a constant loading was assumed as shown in **Table 11** and the yield was predicted. The methane content obtained for both the uniform and non-uniform loadings of RW, MRW and VW is shown in **Figure 11(a)-(c)**. The figures show that the average methane content for simulation and experimental studies is 52% and 53% for RW, 55.69% and 54.85% for MRW and 52.28% and 53.26% for VW respectively.

3.2 Biogas yield prediction for various categories

The pilot study shows that the theoretical and experimental results are similar as shown in **Figure 12(a)**. Therefore, the current approach could be followed for

Phases	Description	Student population	Days
Phase I	Spring working days	1000–2400	1–150
Phase II	Summer break	200–800	151–225
Phase III	Autumn working days	1000–2400	226–315
Phase IV	Winter break	200–800	316–365

Table 10.
Definition of phases according to academic schedule.

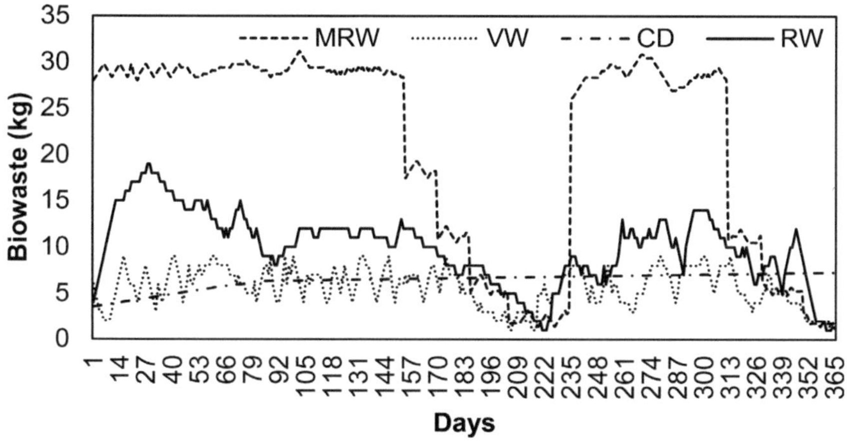

Figure 10.
Loading pattern of biowastes for 365 days.

Biowaste	Biogas yield (m^3)				
	Non-uniform loading				Uniform loading
	Phase I	Phase II	Phase III	Phase IV	Phase (I – IV)
RW	0.16–0.18	0.01–0.05	0.03–0.15	0.01–0.08	0.09
MRW	1.8	0.1	0.8–1.3	0.1–0.3	1.07
VW	0.01–0.18	0.01–0.03	0.04–0.14	0.01–0.09	0.09

Table 11.
Biogas yield during various phases according to academic schedules.

forecasting the biogas yield for different loading rates as shown in **Figure 12(b)**. The yield for each category was determined by academic schedules and biowaste availability.

3.3 Installation and annual operational costs for different biogas plant models

The installation cost and AOC of KVIC, JANATA, and FRP model biogas plants are reviewed for different categories (A to E) as shown in **Figure 13**. The costs of construction, installation, annual service, and other costs are estimated based on the current market price prevailing in the southern part of India.

The Indian government offers subsidies for household digesters regardless of their use. Commercial digesters, on the other hand, are only eligible for subsidies if they are used for power generation. As a result, the subsidy is not considered in this research. The emphasis of the investigation is on the selection of an appropriate biogas plant for non-uniform loading, and its contribution to the reduction of LPG consumption. FRP model has the highest average cost per cubic metre, followed by KVIC and JANATA. The pattern is due to constraints in plant size (12 m^3) and the need for more units. The cost of the KVIC model is higher than JANATA model which may be due to the cost of gas holder. The cost of a gas holder in the KVIC model is high since the steel body needs frequent maintenance; besides, its susceptibility to corrosion. The investment cost is high even though the same gas holder is replaced with FRP. However, the cost of installation for KVIC model decreases steadily from category A to category E, whereas the cost of installation for JANATA

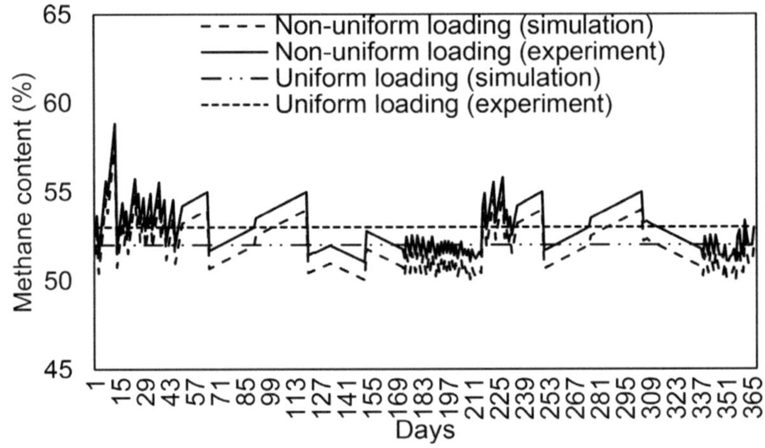

(a).

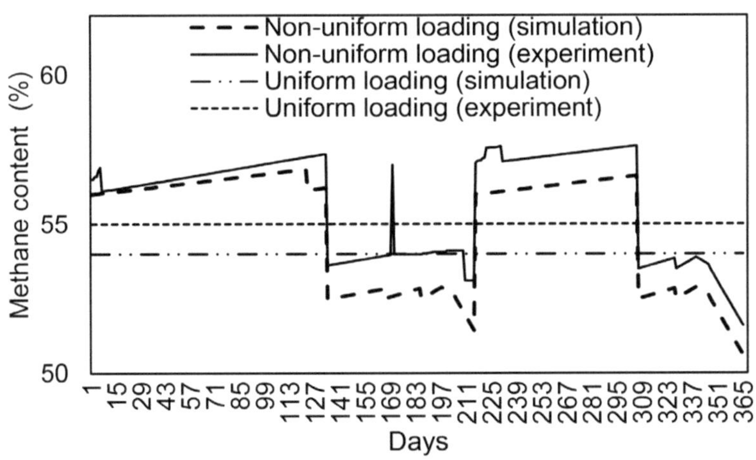

(b).

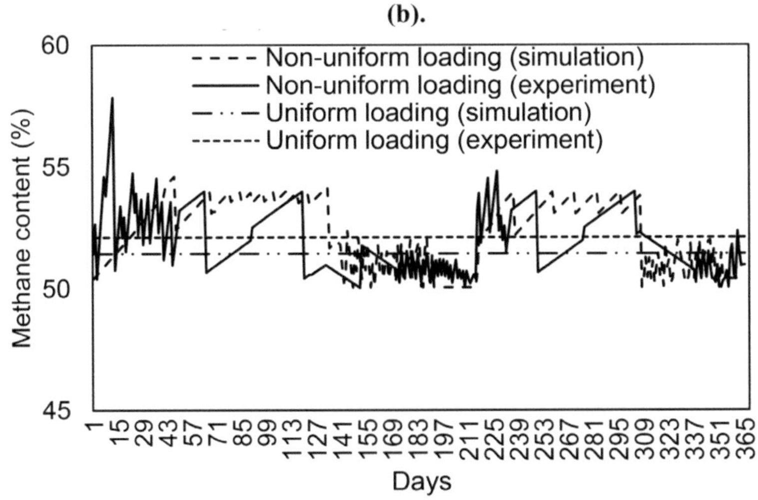

(c).

Figure 11.
(a) Methane content in biogas for rice waste. (b) Methane content in biogas for mixed rice waste. (c) Methane content in biogas for vegetable waste.

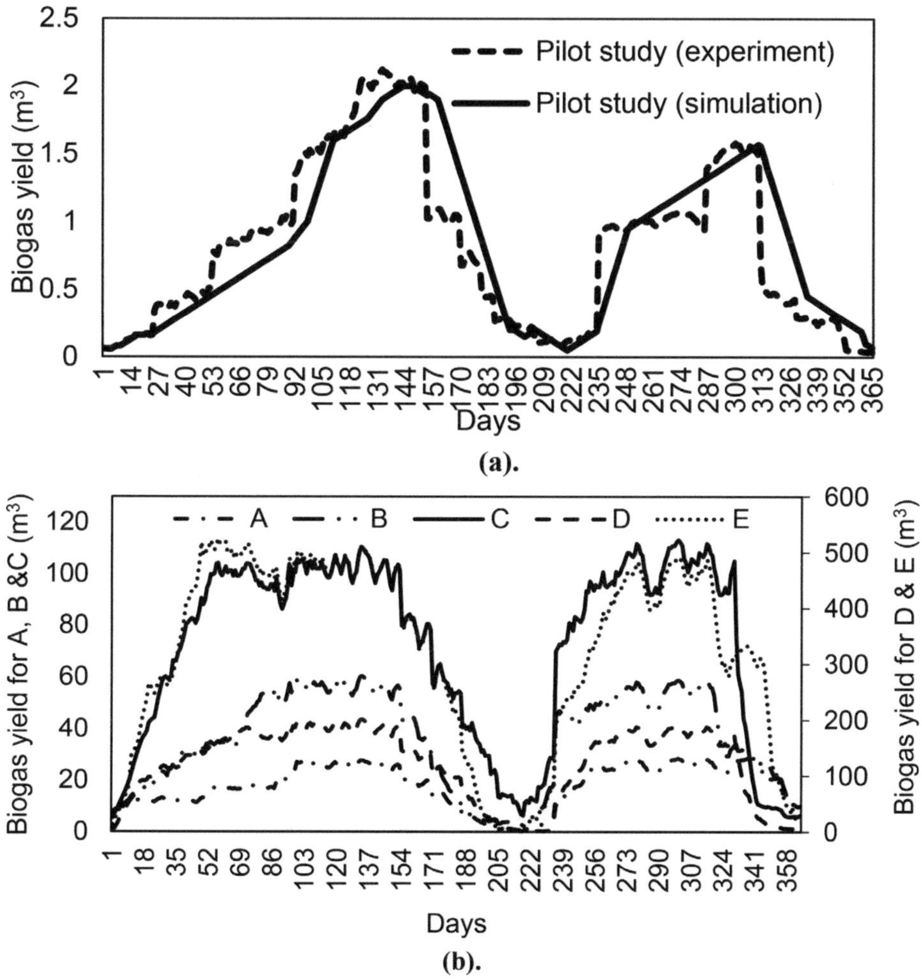

Figure 12.
(a) Biogas yield of pilot plant for 365 days. (b) 365-day biogas yield for categories A, B, C, D, and E.

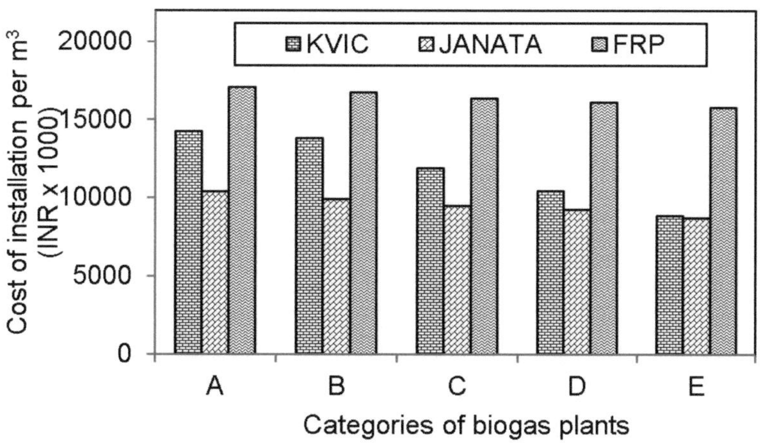

Figure 13.
Installation cost per cubic metre of various biogas plant models.

model is almost same for both categories. **Figure 14** depicts the annual operating cost per cubic metre capacity of all biogas plants in each segment. The FRP model seems to have the highest operating costs, followed by KVIC and JANATA models. The running cost per cubic metre volume for both groups is almost the same for corresponding types and capacities.

3.4 Payback period

The payback period (PBP) of all digesters in various categories has been investigated and is depicted in **Figure 15**. The study reveals that as the volume of the biogas plant increases, PBP decreases, which is consistent with many research findings [45]. The FRP model demands the largest PBP for all categories ranging from 25 to 450 m^3 due to its high construction and operating costs. The KVIC models are well-known for being the most optimal for the production of biogas plants of any size. Though the JANATA style biogas plants are more difficult to build than the other two types, they are very feasible in educational institutions. The payback period for a system with non-uniform loading is 44 to 57 percent

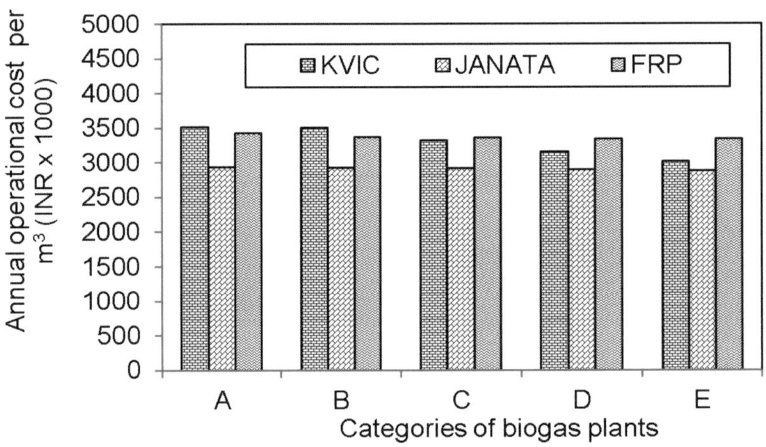

Figure 14.
Annual operational cost per cubic metre of various biogas plant models.

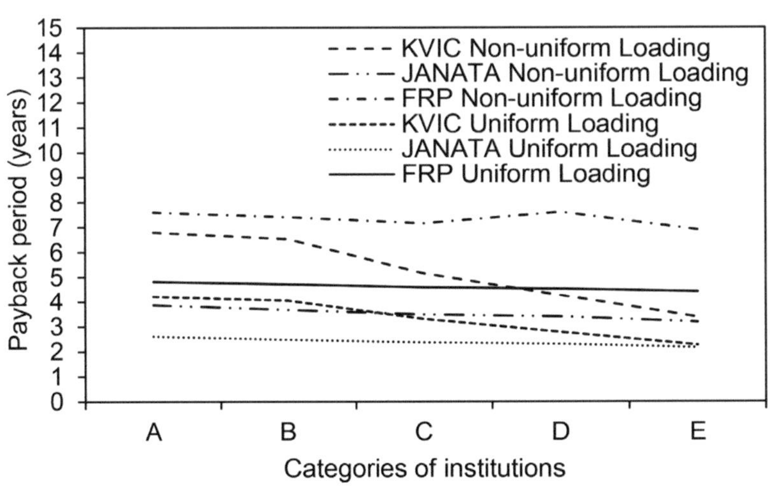

Figure 15.
Payback period for biogas plants for all categories.

longer than for a system that is fully loaded during the year. As a result, if the design and development process is carried out by an expert, the installation of JANATA biogas digester in educational institutions is highly feasible.

3.5 Net present value

The net present value of installing biogas digesters in different types of institutions has been estimated and shown in **Figure 16**.

The NPV of an investment is the difference between the present value of the gains and the present value of the costs arising from the investment. The NPV increases as the scale of the biogas plants increases. The biogas plant project could be preferable for implementation in academic institutions based on NPV selection criteria. The results show that the uniformity in loading produces more useful data than non-uniform loading. However, non-uniform loading rate values indicate that those digesters could be effectively applied in institutions with differing academic schedules.

3.6 Life cycle cost

The most cost-effective solution among competing alternatives that are equally suitable for deployment on technical grounds is determined by a LCC study. As a result, the LCC for uniform and non-uniform loading rates was measured and plotted in **Figure 17**, demonstrating that the LCC of JANATA is the most preferred alternative when compared to the other two versions. However, according to the literature [46], KVIC is recommended because the design and development of larger JANATA model biogas plants is difficult.

3.7 Cost per unit of electricity

The various cost involved in the electricity generation from biowaste available in an educational institution and its equivalent quantity LPG were calculated per year and show in **Figure 18**. The cost of unit electricity was obtained from the following Eq. (8).

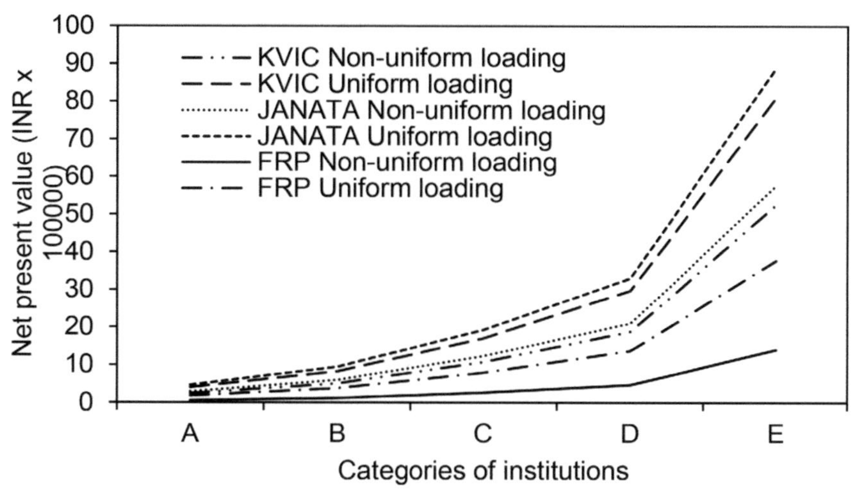

Figure 16.
Net present value of biogas plants for all categories.

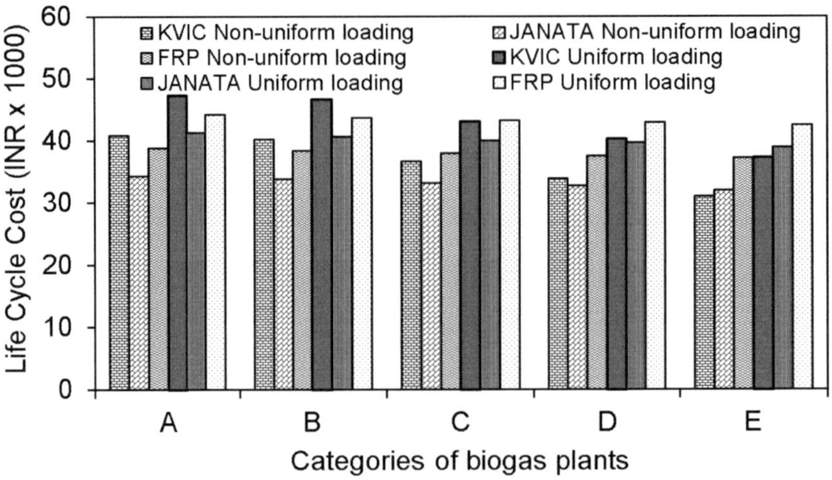

Figure 17.
Lifecycle cost for per cubic meter with uniform and non-uniform loading rates.

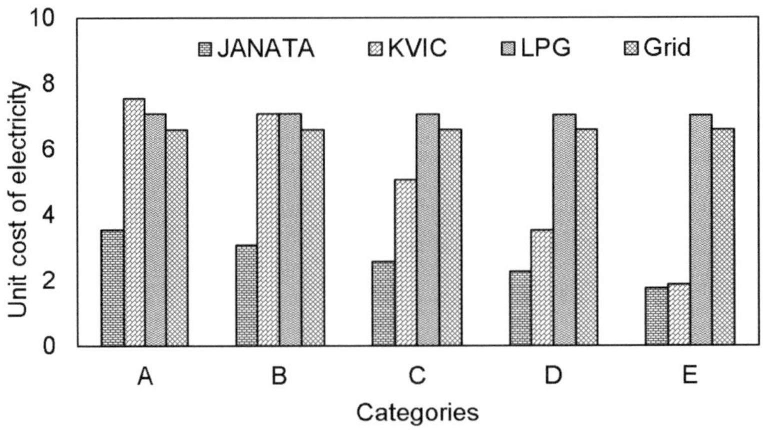

Figure 18.
Comparison of unit cost of electricity from biogas, LPG and grid.

$$\text{Unit cost of electricity generated} = \frac{(Investment\ cost + maintenance\ cost)\ per\ annum}{Total\ units\ generated\ per\ annum} \quad (8)$$

4. Conclusions

The yield of biogas and the efficiency of its production from biowaste of educational institutions, such as rice waste, mixed rice waste, and vegetable waste, were investigated to determine the effect of nonuniform feeding of digesters on the technical and economic viability. As less than 5% of the experimental values were different from the expected content of CH_4 in biogas, the proposed simulation method was found appropriate. Although the biowaste's pH before loading was less than 5, the inoculum's pH was 6.5 to 7.5; thus, the sufficient pH for optimum gas production could be preserved in this method. For all biowastes, the calculated parameters such as total solids, volatile solids and humidity were found within the best suited range of anaerobic digestion. The biogas produced from all biowastes contained 52 to 58% methane which shows that biowastes generated in educational institution included in this study can be used for all types of applications such as electricity generation, lighting and cooling. The amount of biogas generation was affected by population; however, the content of methane in biogas was not affected. In an educational institution, the amount of biogas generated by person per day was 0.014 m^3 to 0.019 m^3 all year. The PBP was 50% higher for both models than that of uniform loading. For the installation in category A, B, C and D institutions based on the PBP, JANATA biogas plants is attractive. JANATA and KVIC are suggested for E group of institutions. The optimistic NPV for the three models and the five separate biogas plant capacities indicates the economic viability of all the designs.

Author details

Godwin Glivin[1*], Mariappan Vairavan[2], Premalatha Manickam[1] and Joseph Sekhar Santhappan[3]

1 Department of Energy and Environment, National Institute of Technology Tiruchirappalli, Tamilnadu, India

2 Department of Mechanical Engineering, National Institute of Technology Tiruchirappalli, Tamilnadu, India

3 Department of Engineering, University of Technology and Applied Sciences, Shinas, Oman

*Address all correspondence to: godwinglivin@gmail.com

References

[1] Uitto JI. Evaluating the environment as a global public good. Evaluation 2016; 22:108–15.

[2] Rasapoor M, Young B, Brar R, Baroutian S. Enhancement of landfill gas generation from aged waste by a combination of moisture adjustment and application of biochar and neutral red additives: A field-scale study. Fuel 2021;283:118932.

[3] Dastjerdi B, Strezov V, Kumar R, He J, Behnia M. Comparative life cycle assessment of system solution scenarios for residual municipal solid waste management in NSW, Australia. Sci Total Environ 2021;767:144355.

[4] Das PK, Rani J, Rawat S, Kumar S. Microalgal Co-cultivation for Biofuel Production and Bioremediation: Current Status and Benefits. BioEnergy Res 2021: 1–26.

[5] Zamri MFMA, Hasmady S, Akhiar A, Ideris F, Shamsuddin AH, Mofijur M, et al. A comprehensive review on anaerobic digestion of organic fraction of municipal solid waste. Renew Sustain Energy Rev 2021;137:110637. https://doi.org/10.1016/j.rser.2020.110637.

[6] Shen R, Jing Y, Feng J, Zhao L, Yao Z, Yu J, et al. Simultaneous carbon dioxide reduction and enhancement of methane production in biogas via anaerobic digestion of cornstalk in continuous stirred-tank reactors: The influences of biochar, environmental parameters, and microorganisms. Bioresour Technol 2021;319:124146.

[7] Hussain F, Shah SZ, Ahmad H, Abubshait SA, Abubshait HA, Laref A, et al. Microalgae an ecofriendly and sustainable wastewater treatment option: Biomass application in biofuel and bio-fertilizer production. A review. Renew Sustain Energy Rev 2021;137: 110603.

[8] Deviram G, Mathimani T, Anto S, Ahamed TS, Ananth DA, Pugazhendhi A. Applications of microalgal and cyanobacterial biomass on a way to safe, cleaner and a sustainable environment. J Clean Prod 2020;253:119770.

[9] Wang R, Li C, Lv N, Pan X, Cai G, Ning J, et al. Deeper insights into effect of activated carbon and nano-zero-valent iron addition on acidogenesis and whole anaerobic digestion. Bioresour Technol 2021;324:124671.

[10] Yang Y, Xu M, Wall JD, Hu Z. Nanosilver impact on methanogenesis and biogas production from municipal solid waste. Waste Manag 2012;32: 816–25.

[11] Fagundes GM, Benetel G, Welter KC, Melo FA, Muir JP, Carriero MM, et al. Tannin as a natural rumen modifier to control methanogenesis in beef cattle in tropical systems: friend or foe to biogas energy production? Res Vet Sci 2020;132:88–96.

[12] Motghare KA, Rathod AP, Wasewar KL, Labhsetwar NK. Comparative study of different waste biomass for energy application. Waste Manag 2016;47:40–5.

[13] Fava F, Totaro G, Diels L, Reis M, Duarte J, Carioca OB, et al. Biowaste biorefinery in Europe: opportunities and research & development needs. N Biotechnol 2015;32:100–8.

[14] Kadam R, Panwar NL. Recent advancement in biogas enrichment and its applications. Renew Sustain Energy Rev 2017;73:892–903.

[15] Nand K, Devi SS, Viswanath P, Deepak S, Sarada R. Anaerobic digestion of canteen wastes for Biogas production: process optimisation. Process Biochem 1991;26:1–5.

[16] Bouallagui H, Cheikh R Ben, Marouani L, Hamdi M. Mesophilic biogas production from fruit and vegetable waste in a tubular digester. Bioresour Technol 2003;86:85–9.

[17] Hernández M, Rodríguez M. Hydrogen production by anaerobic digestion of pig manure: effect of operating conditions. Renew Energy 2013;53:187–92.

[18] Aramrueang N, Rapport J, Zhang R. Effects of hydraulic retention time and organic loading rate on performance and stability of anaerobic digestion of Spirulina platensis. Biosyst Eng 2016; 147:174–82.

[19] Shen F, Yuan H, Pang Y, Chen S, Zhu B, Zou D, et al. Performances of anaerobic co-digestion of fruit & vegetable waste (FVW) and food waste (FW): single-phase vs. two-phase. Bioresour Technol 2013;144:80–5.

[20] Zhang C, Su H, Tan T. Batch and semi-continuous anaerobic digestion of food waste in a dual solid–liquid system. Bioresour Technol 2013;145:10–6.

[21] Schoen MA, Sperl D, Gadermaier M, Goberna M, Franke-Whittle I, Insam H, et al. Population dynamics at digester overload conditions. Bioresour Technol 2009;100:5648–55.

[22] Han W, Fang J, Liu Z, Tang J. Techno-economic evaluation of a combined bioprocess for fermentative hydrogen production from food waste. Bioresour Technol 2016;202: 107–12.

[23] Glivin G, Sekhar SJ. Waste potential, barriers and economic benefits of implementing different models of biogas plants in a few Indian educational institutions. BioEnergy Res 2020;13:668–82.

[24] Glivin G, Sekhar SJ. Studies on the feasibility of producing biogas from rice waste. Rom Biotechnol Lett 2019;24: 728–35.

[25] Glivin G, Sekhar J. Simulation of anaerobic digesters for the non-uniform loading of biowaste generated from an educational institution. Lat Am Appl Res Int J 2020;50:33–40.

[26] Gaul M. A comparative study of small-scale rural energy service pathways for lighting, cooking and mechanical power. Appl Energy 2013; 101:376–92.

[27] Wilkie AC, Graunke RE, Cornejo C. Food waste auditing at three Florida schools. Sustainability 2015;7:1370–87.

[28] Walter WG. Standard methods for the examination of water and wastewater 1961.

[29] Miah MR, Rahman AKML, Akanda MR, Pulak A, Rouf MA. Production of biogas from poultry litter mixed with the co-substrate cow dung. J Taibah Univ Sci 2016;10: 497–504.

[30] Dhanariya R, Sharma S, Sharma KA, Verma S. A review on biogas production from food waste. Int J Pharm Chem Sci 2014;3:899–904.

[31] Ramzan N, Naveed S, Latif N, Saleemi AR. Characterization of kitchen waste as a feedstock for biogas generation by thermophilic anaerobic digestion. NUST J Eng Sci 2010;3:15–21.

[32] Mishra S, Tenneti S. Effect of operational parameters on biogas production using tomato waste as substrate and cow dung as inoculating medium. Int J Sci Res 2015;4:148–52.

[33] Somashekar RK, Verma R, Naik MA. Potential of Biogas Production from Food Waste in a Uniquely Designed Reactor under Lab Condition. 1st IWWG-ARB Symp. Hokkaido Univ. Japan, 2013, p. 18–21.

[34] Abebe M. Characterisation peal of fruit and leaf of vegetable waste with cow dung for maximizing the biogas yield. Int J Energy Power Eng 2017;6: 13–21.

[35] Bond T, Templeton MR. History and future of domestic biogas plants in the developing world. Energy Sustain Dev 2011;15:347–54.

[36] Das B. ROLE OF KHADI AND VILLAGE INDUSTRIES COMMISSION (KVIC) AND RURAL INDUSTRIES-A REVIEW OF LITERATURE. Editor Board 2014;3:102.

[37] Singh KJ, Sooch SS. Comparative study of economics of different models of family size biogas plants for state of Punjab, India. Energy Convers Manag 2004;45:1329–41.

[38] Glivin G, Sekhar SJ. Experimental and analytical studies on the utilization of biowastes available in an educational institution in India. Sustainability 2016; 8:1128.

[39] Glivin G. Techno-Economic Studies on Implementation of Biogas Plant for the Energy Requirement in an Educational Institution. Proc. Int. Conf. Energy Effic. Technol. Sustain. St. Xavier's Cathol. Coll. Eng. TamilNadu, India. 5th to 7th April, 2018.

[40] Rajendran K, Aslanzadeh S, Johansson F, Taherzadeh MJ. Experimental and economical evaluation of a novel biogas digester. Energy Convers Manag 2013;74:183–91.

[41] Saravanan R, Murugavel V. Life Cycle cost Analysis of Waste Heat Operated Absorption Cooling Systems for Building HVAC Applications 2010.

[42] Rajendran K, Kankanala HR, Martinsson R, Taherzadeh MJ. Uncertainty over techno-economic potentials of biogas from municipal solid waste (MSW): A case study on an industrial process. Appl Energy 2014; 125:84–92.

[43] Tiwari GN, Usmani JA, Chandra A. Determination of period for biogas production. Energy Convers Manag 1996;37:199–203.

[44] Kalia AK, Kanwar SS. Long-term evaluation of a fixed dome Janata biogas plant in hilly conditions. Bioresour Technol 1998;65:61–3.

[45] Glivin G, Edwin M, Sekhar SJ. Techno-economic studies on the influences of nonuniform feeding in the biogas plants of educational institutions. Environ Prog Sustain Energy 2018;37: 2156–64.

[46] Kalia AK, Singh SP. Case study of 85 m3 floating drum biogas plant under hilly conditions. Energy Convers Manag 1999;40:693–702.

Chapter 6

Innovative Designs in Household Biogas Digester in Built Neighbourhoods

Isaac Mbir Bryant and Martha Osei-Marfo

Abstract

Most household biogas digesters operate on continuous automatic stirring modes. Often, these digesters rely on electrical energy for their continuous operations which are often mesophilic. Rarely do manually-stirred discontinuous household biogas digesters operating on hyper-thermophilic conditions exist. This work seeks to highlight some innovative designs in a household biogas digester piloted in Terterkessim slum in the K.E.E.A. Municipality of the Central Region, Ghana. A pyramidal dome-shape biogas digester was constructed on an abandoned septic tank using blocks and concrete. The digester has a rectangular sub-surface base and a pyramidal gas holder above the surface of the soil. The digester has a two-blade manual stirrer, a ball bearing affixed at the bottom and a handle to manually mix the content of the digester. In order to heat the content of the digester to a hyper-thermophilic condition for hygienising the digestate, a solar-photovoltaic was installed on the roof of a toilet connected to the household biogas digester.

Keywords: Solar photovoltaic, manual stirrer, hyper-thermophilic, household, biogas digester

1. Introduction

In Sub-Saharan Africa and especially Ghana, the use of renewable energy such as biogas is highly under-developed [1] thus accounting for the country's over-reliance on natural gas and other fossil-based fuels for electrical power generation [1]. It is, therefore, very crucial for Ghana to expand the production of renewable energy such as biogas from food wastes, black water (BW) (waste water comprising human faeces, urine and flush water) for both industrial and household consumption. Consequently, coming up with an innovative and good technological design for household biogas production is very imperative. The choice of the type of reactor and the innovative designs that can be made for efficient technological processes of a household biogas digester in a built is crucial. This is because of the financial repercussions for the citizens (for example, affordability) and its technical complexity for operation and maintenance. In addition, the efficiency and the applicability to the populace especially, in a developing country like Ghana are some of the reasons the choice of a particular innovative design cannot be overlooked. In Ghana, different energy mix is used for various applications such as domestic/residential, non-residential and other industrial facilities (**Figure 1**) [2]. The greatest

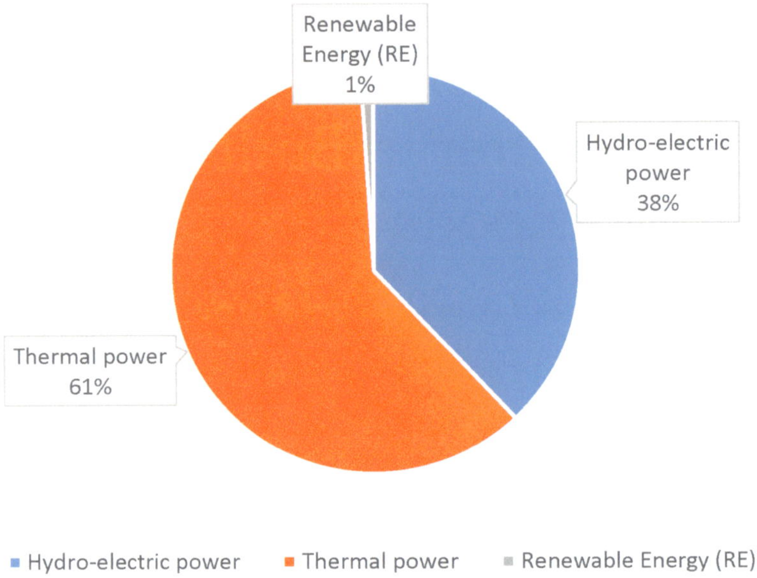

Figure 1.
Percentage contribution of different energy sources used in Ghana. NB: Renewable energy (RE) in Ghana comprises solar energy, energy from biogas, wind energy and biomass energy.

percentage of the energy generation in Ghana is from thermal energy source (61%), followed by hydro-electric power (38%) and 1% making up renewable energy sources such as biogas, solar energy, wind energy and biomass [2].

Different treatment technologies such as Membrane Bioreactor (MBR), Anaerobic Membrane Bioreactors (AMBRs), advanced fluidized bed (AFB) reactors, EGSB and IC® [3] and UASB reactor [4], continuous stirred tank reactor (CSTR) [5] fixed-dome biogas digester (Deenbandhu type) [6, 7] have already been used for biogas production using different substrates and treatment parameters. However, most of these digesters, even though may be modern, did not incorporate other innovative designs that will make them affordable, less technically complex, efficient and easily applicable. This work seeks to address some of these innovative technological missing gaps for easy adoption and implementation, especially, by households in tropical developing countries.

However, single-stage systems are considered to be simple, easy to design and less expensive to be constructed and operated making them common in the anaerobic treatment technology applications [8, 9] Considering small scale anaerobic treatment systems, single-stage reactors have been often used compared to large scale reactors (with a capacity of more than 50 000 tons/year) that use multi-stage systems [7]. According to [7], a fixed-dome (Deenbandhu type) is a closed-dome shaped digester which has an immovable rigid gas-holder. It has an influent inlet and a displacement pit called the compensation tank where the effluent and the digestate exit the reactor. The gas holder is designed to be on top of the digestate in the reactor. With a closed gas valve, higher production of biogas could cause a displacement of the digestate into the compensation tank [6, 7].

The choice of a fixed-dome biogas digester plant for the pilot-scale study in this research for the treatment of household BW in Terterkessim slum in Elmina - Ghana, is based on the following reasons: the user interface is directly connected to the biogas digester [6], the digester can work with or without urine, the reactor can be built underground protecting it from temperature variations [7] and also implies little space is required (making it feasible in a densely populated area like a slum) [6]. Other advantages include: the reactor functions on a wide range of organic input such as animal manure, kitchen waste and BW. Thus, co-digestion would be done to enhance biogas production. It also supports pour flush toilet system (less water used – concentrated BW, higher biogas production), surrounding soil help to counter the in-built pressure in the reactor, moderately not expensive (the use of local materials and labour), has a life span of between 15 to 20 years as there is no corrosion [7].

2. Location for the construction of household biogas digester

The household biogas digester was constructed in Terterkessim slum in Elmina, a coastal town and the administrative capital of the Komenda Edina Eguafo Abirem (K.E.E.A.) Municipality of the Central Region of Ghana [10]. Elmina is bordered to the South by the Gulf of Guinea, West by Bantoma, East by Abakam and North by Bronyibima townships [10]. Elmina lies within latitudes 5° 05' North and 5° 60' North and longitudes 1°20' West and 1° 22' West (**Figure 2**). The town is one of the biggest fishing hubs of Ghana and thus, the major occupation in the town is fishing. The presence of Brenya lagoon, which stretches and overflows (during high tides) to Terterkessim slum, has also made some of the inhabitants to be involved in salt production at commercial quantities. Temperatures are generally high with average being 27°C and annual rainfall ranging between 750 mm to 1000 mm. The vegetation are mostly shrubs and grasses [10]. The town has a total population of approximately 34000, of which about 7600 of the Inhabitants live in Terterkessim slum where the household biogas digester was constructed (Personal Communication with Mr. Damptey- K.E.E.A. Municipal Environmental Health Officer, 2016).

The construction of a household biogas digester connected to a household toilet facility was imperative to help curb the issue of open defecation in the slum due to

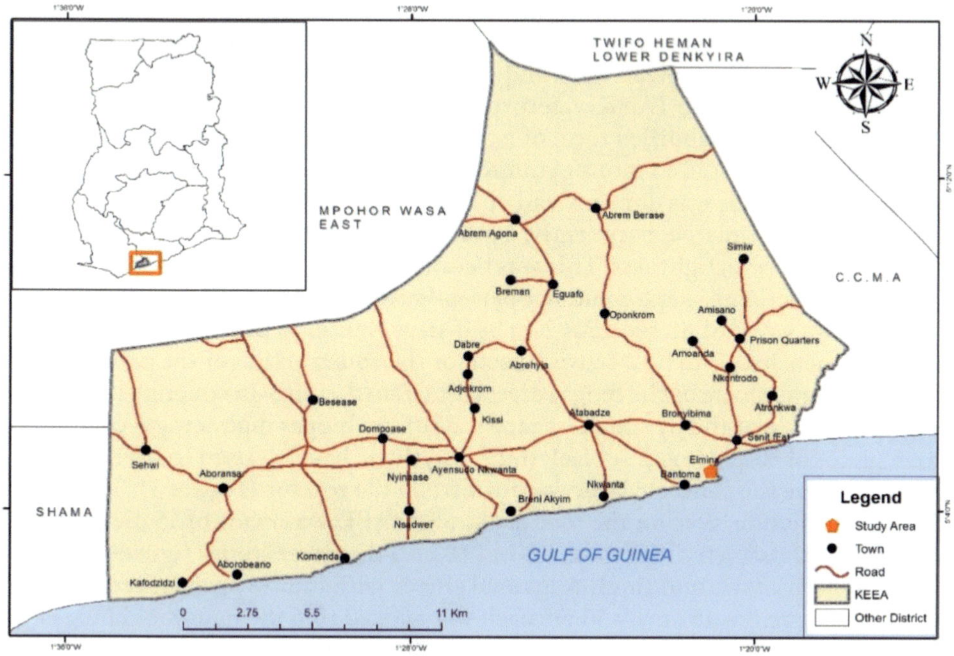

Figure 2.
Map of Ghana showing the district map of the study area, Elmina. Source of the map: Adade, F. (2016). GIS, Department of Fisheries and Aquatic Sciences (DFAS), University of Cape Coast, Cape Coast-Ghana.

lack of public toilets in the community. In addition, the only available toilet facility in the community was in a very bad state. Furthermore, most individual households in the Terterkessim slum do not have household toilet facilities, thus giving the residents the impetus to defecate in the open gutters, lagoon and even in and around the salt ponds. Thus the construction of a household toilet facility connected to a biogas digester with innovative designs for both biogas production and disinfection of digestate was imperative for the Terterkessim urban slum in Elmina.

2.1 Innovative designs for household biogas digesters

The household biogas digester was constructed on an abandoned septic tank thus, it received a lot of modifications to enhance its functionality and efficiency. The innovative designs introduced in the household biogas digester constructed included construction of pyramidal-dome-shape biogas digester, introduction of pour-flush water closet (WC) toilet seats and introduction of manual stirrer in the digester. Other innovative modifications in the household biogas digester built in Terterkessim slum included, adoption of solar photovoltaic to heat the digester to a hyper-thermophilic condition and co-digestion of BW and kitchen food wastes will be highlighted.

2.2 Construction of pyramidal-shape biogas digester

A single-stage household biogas digester was constructed with 6-inch-blocks (moulded sand, cement and water), reinforced with concrete material and plastered with mortar. The concrete material was made of 10 head pans of quarry sand, 10 head pans of 0.5-inch stones (igneous type), 2 bags of rapid strength Portland cement and 10 L of tap water. Additional mortar and water-proof cements like FEB TANK (UK) were used to stop all water leakages into the reactor chambers. The mixture of the mortar was 1 bag of Portland cement (50 kg), 6 head pans of quarry dust, 1 head pan of eroded sand and 2 kg of waterproof FEB TANK cement. About 10 L of water was added and homogenised into a thick paste of mortar for the reinforcement of the weak walls and floor based on the specifications by the manufacturer of the FEB TANK waterproof cement.

The reactor was a modified form of a circular fixed-dome biogas digester with the circular dome modified into a pyramidal-shape roof for biogas storage. The pyramidal shape roof was done instead of the circular dome because the base of the reactor was rectangular, consequently, a pyramidal shape roof on the rectangular base would ensure airtightness. This was because the rectangular base had corners which a circular dome shape could not perfectly fit on without leakages. Ten pieces of 14-ft Wawa wood of dimensions 2-in by 4-in as well as 15 pieces of 14-ft Wawa wood of dimensions 2-in by 2-in were used for the construction of the gable of the pyramidal dome shape of the biogas digester fastened with 3-in concrete nails. The skeletal structure of the pyramidal-shape roof of the biogas digester was covered with 5 pieces of ¼-plywood. A black thick polythene bag was used to cover the plywood before the concrete layer was formed on the reactor (**Figure 3**). The 6-in (15.24 cm) concrete layer for the roof of the SSHTABD was made of 15 pieces of 0.6-in (1.5 cm) diameter iron rods, 1.5-in (3.8 cm) diameter stones (igneous type) and sand (both coarse and fine). A manual stirrer with four (4) galvanised metal blades of dimensions 15 cm by 30 cm each was affixed into the household biogas digester (**Figure 4**). The rotating metal rod of the stirrer was welded into two ball bearings (one affixed to the bottom of the concrete and the other at the top of the metal rod just beneath the pyramidal shape) to enhance easy rotational movement when manually stirred.

Figure 3.
Construction of a pyramidal-shape biogas digester insulated with black polythene bag.

Figure 4.
Manual stirrer in the biogas digester.

2.3 Pour-flush water closet toilet

Two pour-flush water closet (WC) toilet seats were installed in each of the toilet unit connected to a household biogas digester (**Figure 5**). Polyvinyl chloride (PVC) pipes of diameter 4-inches were connected to the toilet seats and into the main chamber of the digester. Adjoining pipes from the WC into the digester were

Figure 5.
A 3-litre pour-flush toilet seat connected to the biogas digester.

connected using 4-inch Tee, 4-inch 45° and 4-inch 90° pipes. The influent pipe was inserted into the reactor to a depth of 450 mm above the floor of the reactor. This was done to ensure that the influent fully covered the pipe to avoid any biogas leakage through the influent pipe. An inlet pipe with a cover was also connected to the influent pipe carrying faecal materials to enhance co-digestion processes (**Figure 6**).

2.4 Manual stirrer

Most biogas digesters that operate in a continuous mode and use stirrers that rely on electrical energy for stirring the digesters. In this design, a manual stirrer was introduced into the household biogas digester for discontinuous stirring by the users in the household. The users were educated and trained to manually stir the digester anytime they visited the toilet. In this way, it was ensured that the old and new feedstock would easily mix to enhance faster digestion. The manual stirrer with four (4) galvanised metal blades of dimensions 15 cm by 30 cm each was affixed into the household biogas digester. The rotating metal rod of the stirrer was welded into two ball bearings (one affixed to the bottom of the concrete floor of the digester and the other at the top of the metal rod just beneath the pyramidal shape) to enhance easy rotational movement when manually stirred (**Figure 4**).

2.5 Installation of solar-photovoltaic for heating the digester

A high quality 50 W offgridtec® autarkic mono photovoltaic panel of dimensions 60.5 cm x 47.5 cm (0.3 m^2) was installed on the roof of the toilet connected to the SSHTABD for heating. The photovoltaic panel was offgrid with model number 3–01-001260 and had a voltage of 22.3 V (made by offgridtec® AGM GmbH, CMK ENERGY, Germany). The photovoltaic panel was connected to a solar charge controller (Stecca PR1010 756.477 by Solar Electronics, PV offGrid, PV Autarke

Figure 6.
Pipe connections from the pour-flush toilet into the innovative household biogas digester.

systeme, made in EU) via solar cables. The charge controller was connected to a 12 V/30 A/20 Hours offgridtec AGM gel battery series (by offgridtec AGM GmbH, Germany). The battery had a constant voltage charge and voltage regulation with cycle use of 14.5–14.9 V at 25°C and standby use of 13.6–13.8 V at 25°C. The battery was connected to an NP series pure sine wave inverter (Model number NP 300, made by Solartronics, Leipzig-Germany) which had a maximum peak power of 600 W and an average current of 300–400 W. It also had an input voltage of 12 V and an output voltage of 230 V - 50 Hz and efficiency of 84–94% (**Figure 7**).

2.6 Installation of galvanised copper pipes into kitchen

Galvanised copper pipes were used to connect the SSHTABD to the kitchen of the household where potential biogas to be produced was to be used. The copper pipes had diameter of 2 cm. Stop corks or valves were installed at adjoining points to regulate the flow of biogas into a biogas bag to monitor the daily biogas production. The copper pipe was laid into the walls of the restroom to the kitchen at an angle of 45° in order to ensure that all water vapour that could form during the operation of the SSHTABD would trickle down by gravity into a collection tube to be discharged (without losing biogas from the reactor) (**Figure 8**).

Figure 7.
Components of solar-photovoltaic system installed on the biogas digester.

Figure 8.
Installation of galvanised copper pipes for tapping biogas into kitchen.

2.7 Detailed description and performance of innovative household biogas digester

The single-stage innovative household biogas digester constructed in Terterkessim slum composed of 3 chambers which were originally designed for a septic tank system. The septic tanks were connected to a two-unit toilet meant for that household. The first chamber was the biggest and was converted into the main single-stage household biogas digester in which the AD process occurred. It had a total volume of 8.64 m^3. Adjoining the main reactor was a compensation tank which had a tunnel from the main digestion chamber. The compensation tank was about 3.17 m^3. Within the compensation tank were steps designed to help with settling of particles as well as directing clear effluent to be discharged into the next chamber, the effluent collection and storage tank. The effluent collection and storage tank had a total volume of 4.52 m^3. It had an effluent discharge pipe for overflow into a collection container for agricultural usage. An average COD removal of 97.6% was recorded for the digester. The operational parameters for the innovative household biogas digester were a mean temperature of 37°C, average daily flow rate of 182.1 L/d and mean HRT of 51.3 days. The mean daily volumetric loading rate and mean daily organic loading rates of 0.97 kgCOD/(m^3.d) and 0.06 kgVS/(m^3.d), respectively, were also recorded for the digester. These operational values for the biogas digester gave an implication the digester had more potential of receiving more organic load for treatment daily. The digester could produce about 2.52 Nm^3CH_4/(kgCOD.d) which could be burnt for at least 8 hours for purposes such as cooking and heating in the households in the slum. This high value was recorded because of the simultaneous conversion of food waste and human faeces into biogas.

3. Conclusions

Manually-stirred discontinuous household biogas digesters which also operate on hyper-thermophilic conditions for anaerobic digestion processes rarely exist. In this study, the objective was to highlight some innovative designs in a household biogas digester piloted in a slum called Terterkessim in the K.E.E.A. Municipality of the Central Region of Ghana. A 2-seater toilet compartment was constructed on a pilot manually-stirred, fixed pyramidal-dome-shaped single-stage household biogas digester for a compound house of 32 persons in the Terterkessim slum. The pyramidal dome-shape biogas digester was constructed on an abandoned septic tank meant to contain faeces from the toilets. Blocks and concrete were used for the construction. The digester has a rectangular sub-surface base and a pyramidal gas holder above the surface of the soil. It also has a two-blade manual stirrer, a ball bearing affixed at the bottom and a handle to manually mix the content of the digester. A solar-photovoltaic was installed on the roof of the toilet connected to the digester to heat the content to a hyper-thermophilic condition for hygienising the digestate.

The innovative household biogas digester has a potential to produce about 2.52 $Nm^3CH_4/(kgCOD.d)$ which could be burnt for at least 8 hours for purposes such as cooking and heating in the household. With average daily flow rate of 182.1 L/d and mean HRT of 51.3 days, 97% of the influent was removed. Consequently, this innovative household biogas digester can be employed in already existing residential facilities or new residences for wastewater treatment at the household level and energy recovery from the waste.

Acknowledgements

We wish to acknowledge the contributions of all laboratory and technical staff of the School of Agriculture laboratory in the University of Cape Coast Technology Village for their immense contributions during the laboratory analyses.

Conflict of interest

"The authors declare no conflict of interest."

Notes/thanks/other declarations

We declare that the information provided in this Book Chapter form part of the data and work obtained by the corresponding author of this scholarly work and needs no copy right declaration. In addition, the authors wish to declare that the research was not funded by any organisation.

We wish to thank all labourers and artisans who supported with the construction of the household biogas digester.

Author details

Isaac Mbir Bryant[1]* and Martha Osei-Marfo[2]

1 Department of Environmental Science, School of Biological Sciences, University of Cape Coast, Cape Coast, Ghana

2 Department of Water and Sanitation, School of Physical Sciences, University of Cape Coast, Cape Coast, Ghana

*Address all correspondence to: ibryant@ucc.edu.gh

References

[1] Kumi, E. N. (2017). The Electricity Situation in Ghana: Challenges and Opportunities www.cgdev.org, (September), 30. Retrieved from www.cgdev.org

[2] Ghana Renewable Energy Master Plan (2019).

[3] van Lier, J. B. (2008). High-rate anaerobic wastewater treatment: diversifying from end-of-the-pipe treatment to resource-oriented conversion techniques. Water Science & Technology, 57(8), 1137. https://doi.org/10.2166/wst.2008.040

[4] Kujawa-Roeleveld, K., & Zeeman, G. (2006). Anaerobic Treatment in Decentralised and Source-Separation-Based Sanitation Concepts. Reviews in Environmental Science and Bio/Technology, 5(1), 115-139. https://doi.org/10.1007/s11157-005-5789-9

[5] Tchobanoglous, George; Burton, Franklin L. and Stensel, H. D. (2004). *Wastewater Engineering: Treatment and Reuse*. *Metcalf & Eddy, Inc.* (Fourth Edi). Metcalf & Eddy, Inc. https://doi.org/10.1007/s13398-014-0173-7.2

[6] Tilley, E., Lüthi, C., Morel, A., Zurbrügg, C., & Schertenleib, R. (2014). Compendium of Sanitation Systems and Technologies. Development, 180. https://doi.org/SAN-12

[7] Vögeli, Y., Riu, C., Gallardo, A., Diener, S., & Zurbrügg, C. (2014). *Anaerobic Digestion of Biowaste in Developing Countries*. Retrieved from http://www.eawag.ch/forschung/sandec/publikationen/swm/dl/biowaste.pdf

[8] Foley, T., Thornton, K., Hinrichs-rahlwes, R., Sawyer, S., Sander, M., Taylor, R., ... Hales, D. (2015). Renewables 2015 Global Status Report. Retrieved from http://www.ren21.net/wp-content/uploads/2015/07/REN12-GSR2015_Onlinebook_low_nolinks.pdf

[9] Rapport, J., Zhang, R., Jenkins, B. M., & Williams, R. B. (2008). Current Anaerobic Digestion Technologies Used for Treatment of Municipal Organic Solid Waste, (March), 69 p. https://doi.org/10.1016/j.biortech.2006.11.048

[10] Ghana Statistical Service, (2014). Komenda-Edina-Eguafo-Abrem Municipal. *2010 Population and Housing Census, K.E.E.A. Municipality*, 1-87.